trotman **education**

The only guides to compare th... structur...

CW00602474

Physics & Chemistry

9 SERIES 1

CRAC
Degree Course Guides 2007/08

The CRAC Series of Degree Course Guides

The CRAC Series of Degree Course Guides
Physics and Chemistry

Compilers and series editors
Andrew Smith and Marion Owen

This Guide published in 2007 by Trotman, an imprint of Crimson Publishing, Westminster House, Kew Road, Richmond, Surrey TW9 2ND
www.crimsonpublishing.co.uk

British Library Cataloguing in Publication Data
A catalogue record for this book is available from the British Library

ISBN 978-1-90604-116-8

Printed and bound in Great Britain by Bell & Bain, Glasgow

This Degree Course Guide is one of a series providing comparative information about first-degree courses offered in the UK. The aim of the series is to help you choose the course most suitable for you.

This Guide is not an official publication, and although every effort has been made to ensure accuracy, the publishers can accept no responsibility for errors or omissions. Changes are continually being made and details have not been given for all courses, so you must also consult up-to-date prospectuses before finally deciding which courses to apply for. The publishers wish to thank the institutions of higher education that have provided information for this edition.

CRAC: The Careers Research and Advisory Centre

CRAC is an independent not-for-profit organisation. CRAC Degree Course Guides are published under exclusive licence by Trotman Publishing. For catalogues or information, write to Trotman Publishing; to buy publications, contact NBN International, Estover Road, Plymouth PL6 7PY. Tel: 0870 900 2665.

Physics and Chemistry

Academic consultants:

Physics
Professor Peter Main,
Institute of Physics

Chemistry
Dr Josephine Tunney,
Royal Society of Chemistry

Foreword

With the recent changes to student funding and tuition fees, entering higher education can no longer be a decision made lightly by school-leavers. Though overall there has been an increase in the numbers applying to university in the last five years, those numbers have fluctuated and the funding changes have muddied the waters for students who do not have the ready funds to enter, and leave, higher education debt-free.

So is it still worth it? There is evidence that there is still a 'graduate advantage'. Salaries for graduates across the span of a career remain higher than for non-graduates. The demand in the UK for higher skills is great; it is clear that higher levels of skills are associated with higher levels of productivity, and greater productivity coupled with higher-level skills are vital to the growth of the UK economy.

Career choices for graduates are constantly changing. Changes in industry mean that there is an increasing demand for graduates to enter marketing, IT, management accountancy, management consultancy, community and society occupations and other newer professions. Graduate career paths are not necessarily clear-cut, but the majority of graduates find themselves in employment that is related to their long-term career plans. One growing trend is that employers are more and more interested in work experience. The good news here is that universities are increasingly catering for this by providing assessed placements that form part of the degree course and opportunities for volunteering in the local community.

Students and graduates feel that their decision to enter higher education is worthwhile. The 2006 National Student Survey, the official survey of students' feelings about their courses, shows that approximately 80% were satisfied with their higher education compared to their expectations. According to the research project *Seven Years On: Graduate Careers in a Changing Labour Market*, more than two-thirds of graduates say they would do it all over again.

Thus it seems that the prospects for new entrants to higher education are still exciting. Making this decision may be daunting, bewildering or even liberating. It is vital that you take into account all of the factors that you feel are important in making a choice: your career aspirations, your abilities and aptitudes, the teaching facilities, opportunities for placements and, importantly, your passion for the subject. The choice of institutions and degree programmes is vast, and creating a shortlist can prove to be challenging – the biggest challenge being to ensure your choice of course and institution is an informed one. The CRAC *Degree Course Guides* are intended to be accurate, comprehensive and insightful to help equip you to do just this, making them essential reading for anyone considering entry to higher education.

Jeffrey Defries, Chief Executive
The Careers Research and Advisory Centre (CRAC)
www.crac.org.uk

The tables provide comparative information about first-degree courses offered at higher education institutions (universities and colleges) in the UK. Each table provides detailed information organised alphabetically by institution and course title. An at-a-glance summary of the material in each main table is given below:

Table 2a: First-degree courses

This gives the degree qualification, duration, foundation year availability, modes of study, whether the course is part of a modular scheme, course type and number of combined courses available.

Table 2b: Subjects in combination

This lists subjects that can be combined with physics or chemistry, and where they can be studied.

Tables 3a–3d: Course content

These give detailed information on the content and organisation of the courses, and special features such as study abroad.

Table 4: Assessment methods

This gives the frequency of assessment, years in which there are exams contributing to the final degree mark and percentage assessed by coursework.

Table 5: Entrance requirements

This gives the number of students on each course, typical offer in UCAS points, A-level grades and Scottish Highers, and required and preferred subjects at A-level.

'How to use this Guide' provides an overview of the book's content and more detailed information is provided before each table.

How to use this Guide

This Guide provides you with comparative information about the honours degree courses in physics and chemistry that you can take at higher education institutions in the UK. It aims to help you answer some of the questions you need to ask if you are going to find the courses that suit you best and interest you most. Some of these questions are shown below, with an indication of where you should look in the Guide for answers.

Physics courses appear in the first part of this Guide, chemistry courses in the second: the chapters are numbered from 1 in each part, so references to specific chapters in the outline below refer to the relevant chapters in each part of the Guide.

What are these subjects like at degree level? The first chapter is designed to give you some insight into what studying physics or chemistry is like at degree level. It gives an indication of what you will learn and how you will learn it, with an overview of the very wide variety of courses on offer. The chapter also gives you some idea of the careers opportunities open to graduates in these subjects.

What courses are offered in physics and chemistry? First you will need to decide whether a specialised or combined course will suit you best, and how long you want to spend on the course. You can get basic information on these points from TABLE 2a, which lists all courses in which one of these subjects can make up at least half of a degree course. TABLE 2b then shows which subjects you can combine with physics or chemistry in a two-subject degree. At the end of Chapter 2 there is a list of courses in related areas, which may interest you as an alternative to a physics or chemistry course.

What are the courses like? One of your major concerns will naturally be to find out just what you will study during your undergraduate career. Chapter 3 describes the style and content of the courses, mainly using tables so that you can compare courses easily. Before you dive into the tables, you should read the surrounding text, as this will tell you how to interpret them and what they can and cannot tell you.

How will my work be assessed? Institutions in higher education use a wide variety of assessment methods, and the balance between them is often different at different institutions. Chapter 4 identifies and distinguishes between the approaches used.

What do I need to get accepted for a course? Chapter 5 describes the qualifications you will need for admission to degree courses covered by this Guide, and TABLE 5 lists the requirements for individual courses and gives an indication of the number of places on offer. There is no simple relationship between the number of places on offer and how easy it is to get accepted for a course, but this figure does at least give you some idea of how many fellow students you would have on a course.

What professional organisations support physicists and chemists? Chapter 6 gives information about the professional organisations relevant for physicists and chemists. These do not play quite the same role as, for example, the professional bodies in medicine and law, which have statutory responsibility for governing the profession. Nor do they have the same practical influence as the engineering institutions, which under the Engineering Council exert very strong control over the education and professional development of engineers. However, they are influential in their fields and provide many valuable services for students and professionals. The Institute of Physics can also award chartered status after you have satisfactorily completed your degree and a period of recognised work experience.

Where do I go from here? This Guide, like the others in the series, aims to provide comparative information about courses to help you decide which ones you want to follow up in more detail, but it is only concerned with the content and organisation of the courses themselves. You will need to look at prospectuses, websites such as www.trotman.co.uk and the institutions' own sites, shown in TABLE 2a, to find out about such things as tuition fees, accommodation, locality and student life at each institution. Chapter 7 gives a list of sources of further information, including a contact name for each course. In addition, most institutions welcome visits from potential students, and many run open days. Once you have applied, you may be invited for an interview, which will usually include an opportunity to look around the institution's facilities.

Part 1

Chapter 1: Introduction

Physics is a very broad subject covering a vast range of topics and suiting a wide variety of interests and temperaments. The subject is so large and highly developed that many institutions offer extended MPhys or MSci courses to increase coverage and to allow time for the teaching of skills such as IT and communications: see Chapters 2 and 3 for details. These last a year longer than most degrees and allow you to reach the level needed to practise as a physicist in industry or research.

The scope of physics Physics deals with phenomena ranging in size from the entire universe to what is known as the Planck length (10^{-35}m) – to give you an idea of what this length means, the size of the nucleus is roughly intermediate between the Planck length and the size of the solar system. Physicists have explanations for the way the universe has evolved from one-hundredth of a second after it was created, and reasonable speculations about what happened at even earlier times. They can also predict how the universe might end, though the evidence at present is tantalisingly ambiguous about whether the universe will come to an end or carry on indefinitely.

Experimental physicists have obtained temperatures from close to absolute zero to higher than those found in the Sun. The fact that we know there is such a thing as absolute zero was in itself one of the major triumphs of classical physics. There are theories dealing with matter as dense as the nucleus and as tenuous as intergalactic space, and with radiation of all wavelengths from the longest radio waves to the highest energy gamma rays. You may come to see some theories as little more than formalised common sense, but others as stranger than the most bizarre science fiction. There is a place in physics for the most hands-on experimentalist working in industry and for the most abstract theorist contemplating whether space and time has 4, 10 or 26 dimensions.

Precise and accurate Perhaps one of the most striking features of physics is the extraordinary precision and accuracy of its predictions. There are branches of the subject where you count yourself successful if you get agreement between experiment and theory within one or two orders of magnitude, but there are others where agreement is better than one part in ten million million. No other science has produced theories that have been tested as rigorously as those in physics. For example, Newton's laws and a few other parts of classical physics are the basis of most engineering calculations, which are tested every day by the fact that machines, cars, aeroplanes, in fact anything that moves, behave as expected, or, if they don't, the same laws are used to explain why. And this is despite the fact that Newtonian mechanics is known to be inaccurate and has been replaced by Einstein's theory of relativity: even an inaccurate and superseded law of physics can be good enough for practical purposes.

The past and the future Physics is the oldest science, dating back thousands of years to the Ancient Greeks. Despite this, physicists have more problems to investigate and challenges to face than ever before. As more progress is made, new explanations are required. Improvements in theory and technology lead to new phenomena being discovered, and new possibilities for testing theories and showing up discrepancies that need to be explained.

You may have heard of Stephen Hawking's prediction that we are not far away from an ultimate theory of everything. Even if he is right, and it is not clear that he is, all such a theory would do is explain the basic structure of space and time, which determines the fundamental particles and forces making up the universe. Of course, this would be a major achievement, but of absolutely no help in tackling the problems addressed by the majority of physicists, who have to deal with the limitations of technology and work in areas dealing with very complex systems of interactions, such as solid-state or low temperature physics.

Physics is useful Having looked at some of the more esoteric boundaries of the subject, you may have the impression that physics is entirely theoretical and academic. Nothing could be further from the truth. Physics is the basis of nearly all engineering disciplines: even chemical engineering, which you might think was more closely related to chemistry, relies on physics for an understanding of thermodynamics, which controls the speed of reactions, and for calculations of heat and fluid flow.

Materials and electronics Industry uses an enormous variety of materials for its products and during the production process. Physics is fundamental to an understanding of the properties of all sorts of materials. For example, the behaviour of electronic devices can only be understood and predicted with a knowledge of semiconductor physics. Physicists are involved in the search for new semiconducting materials with special properties allowing higher performance, and in the development of new fabrication techniques such as X-ray and particle beam lithography, to achieve the much higher component densities on chips that will be necessary if the computing industry is to continue its trend of doubling power and halving cost every few years. Most recently, physicists have developed apparatus that can image and manipulate matter at the atomic scale, which has led to a whole new research area under the broad headings of nanoscience and nanotechnology.

Lasers and superconductivity For many years lasers were an academic curiosity in the physics laboratory, but they are now found in the scanners at supermarket checkouts, in computer printers in the office and in CD players in the home. Superconductivity (some materials have zero electrical resistance at very low temperatures) has been known and understood for a long time, but the conditions required have meant there have been few practical applications. The recent development of high temperature superconductors (high temperature means the temperature of liquid nitrogen rather than liquid helium) has opened up the possibility of much wider use, but a great deal of basic physics research needs to be done before this will be achieved.

Medicine Surgeons use lasers for delicate operations. Radiographers use radioactive isotopes, ultrasound scanners, magnetic resonance imaging, X-ray machines and computed tomography scanners. Intensive care units rely on a battery of monitoring equipment. All of this has been developed through an understanding of basic physics combined with the imagination required to apply that understanding to solve problems in new fields.

From A-level to degree physics Unless you are thinking of taking a foundation course (see Chapter 2), you will need to have studied physics at A-level or an equivalent standard – most universities also require you to have studied mathematics to a similar level (see the *Mathematics* section overleaf for more on this). Physics degree courses start off by building solid foundations in classical physics. Most of this will be in areas familiar from your A-level studies but taken much further and with greater rigour.

Typically, courses will have topics on classical mechanics based on Newton's laws, and on electromagnetism, which will introduce Maxwell's laws. At some stage, possibly even in the first year, you will be introduced to special relativity, which reconciles Newtonian mechanics with Maxwell's laws. You will also have topics on properties of matter, which will build on A-level work on atomic theory and the gas laws, and which may include some thermodynamics. Wave motion is an important link between a number of different branches of physics including electromagnetism, optics and acoustics, and is also usually the basis for work in quantum mechanics, which may also be introduced in the first year.

A feature of physics degree courses that you will recognise from A-level is that work in later years builds directly on earlier work, so you must make sure that you get to grips with it from the beginning.

Later parts of the course will continue to build on the classical foundations, but will also introduce you to more modern physics such as astrophysics and cosmology, plasma physics, quantum theory, nuclear and elementary particle physics, statistical mechanics and low temperature physics.

Teaching style Probably the biggest difference you will notice when you start a degree course in physics is the change in teaching style. At school you will be used to small classes with frequent interaction with the teacher during lessons, and supervision and help in carrying out examples. There is also usually a close integration between theory and practical work. The larger numbers, particularly in the early stages of most physics courses, mean that you have to take much greater control of your own learning. The basic material is presented in lectures, which are usually much less interactive than school classes and, apart from the occasional demonstration, will be mainly theoretical. Practical work is usually related to the work in lectures, but unlike at school, it would be unusual for the theory to be taught through practical work or even for you to be able to follow up a piece of theory work with relevant practicals on the same day. The number of students taking a first-year practical class in a large department may also come as a shock after what may have been a small and intimate A-level practical group. You are usually much more on your own in following up theory work with examples, though most institutions run examples classes and tutorials where your work is supervised and you have the opportunity to get help.

A feature of many courses, and some MPhys and MSci courses in particular, is an emphasis given to the teaching of skills as well as content. IT skills are particularly important, and physics courses all have extensive training in this area.

Most students react positively to the extra freedom they have, but inevitably a minority are unwilling or unable to take on the responsibility for planning and carrying out their own studies. You need to be aware of these demands and make sure that you are prepared to live up to them before you start any degree course.

Mathematics You will already know from your experience at A-level that mathematics is the basic language of physics. Indeed, there is virtually no distinction between theoretical physics and certain branches of applied mathematics. Given this, it will come as no surprise that physics courses have a fairly high mathematical content and will include some lectures in specific areas of applied mathematics. The actual level of the mathematics covered varies from course to course and option to option, but in all courses you will need to be comfortable expressing ideas mathematically. In particular, theoretical options and specialist courses in theoretical physics usually reach a level of mathematics comparable to that found in the applied mathematics part of a mathematics degree course.

What can you do with a physics degree?

Physicists' specialist knowledge is important to many employers, and their problem-solving skills are valued in numerous areas of work. As a result, a wide variety of jobs are open to physics graduates, the most popular areas being in scientific/engineering research and development, and information technology. The following sections give some idea of the many possibilities open to you as a physics graduate.

Further study and training Typically, well over a third of physics graduates go on to some form of further study or training. This figure is roughly twice that for graduates as a whole, reflecting the importance of a higher degree for careers in research. A large majority of those going into further study or training are working towards a PhD or following a Master's course. Working for a PhD allows you to research a specialised topic in great depth and make an original contribution to that field of knowledge. Master's courses enable you to add a further area of specialist knowledge such as optics, nuclear physics or elementary particle theory, and gain some research experience. Courses are also available in related areas such as information technology. Greater opportunities are available in research and development jobs for those with a higher degree, but you will need at least a very good upper second-class degree for a PhD and for many Master's courses.

Another option for further study is a vocational course such as teacher training (see below for more on teaching), though comparatively small numbers of physics graduates follow this route.

Electronics, telecommunications and related industries The electronics industry, computers and telecommunications systems, including laser optics, have all developed from an understanding of the laws of physics. Consequently, there are opportunities for physicists to work on the research, development and design of electronic devices and integrated circuits, magnetic materials, lasers and image processing systems. Employers include manufacturers and service providers, such as BT. For more information on careers in physics, visit the website of the Institute of Physics at www.iop.org.

Materials The study of the properties of materials is an important branch of physics, so it is not surprising that physicists have a lot to offer materials manufacturers and processors, including those in the steel, plastics and ceramics industries, not to mention the enormous variety of companies that use these and other materials to produce finished goods. The physical properties of different materials are studied with techniques such as diffraction and spectroscopy, vibration, hardness testing and many other physical tests. Measurements of the properties of paints or gels, and colour analysis of photographic films, are other areas in which physicists' expertise is critical.

The energy industry, communications and control The energy industry is another important area offering career opportunities for physics graduates. For example, geophysicists work on seismic surveys in the oil exploration and production industry. The electricity generation and supply industry, oil and gas companies, and producers of solar panels all need physicists to optimise their operations, seek more environmentally friendly methods and continue the search for more efficient ways of generating, storing, transmitting and using energy.

Physicists can use their knowledge of classical mechanics and dynamics, together with skills in electronics and computing, in careers working on communications satellites, missile trajectories, turbines and control systems.

Healthcare There are several opportunities for physicists to become involved in healthcare. Hospital physicists work in hospitals calibrating and maintaining equipment as well as using it with patients. The manufacturers of medical appliances also employ physicists to design and develop equipment, calibrate and install it for customers, and in the technical documentation and marketing effort. Examples of the sort of equipment used include sophisticated machines such as ultrasonic scanners for picturing the unborn child, whole-body scanners and magnetic resonance imaging (MRI) techniques, as well as X-ray machines and radioisotopes used for treating tumours.

Teaching in schools Before you can teach in a state school you must first take and pass the one-year Postgraduate Certificate in Education course. There has been a severe shortage of school teachers in many subject areas, including physics, for some years. The government has sought to address this through a series of financial incentives, which currently include – subject to various conditions – a bursary of £9,000

(£7,200 in Wales) while you train, and a 'golden hello' of £5,000 when you enter your second year of teaching in a state school in England or Wales.

Now that the teaching of science in schools tends to be integrated, physics teachers usually work together in teams with biologists and chemists, and may find themselves teaching significant amounts of biology and chemistry in addition to their own subject. It is also possible to train to teach at primary or further education level. For more on teaching, visit the Training and Development Agency for Schools website at www.tda.gov.uk.

Research careers A few highly qualified physicists obtain academic positions in universities. A first step for those aspiring to a university research career is usually to complete a PhD, followed by some years as a postdoctoral researcher. Research assistant positions are another option. These are paid research posts that combine research work with studying for a PhD and often some undergraduate teaching as well. There is great competition for the small number of more permanent academic posts that arise.

The fundamental problems studied in some branches of physics such as astronomy, elementary particle physics and nuclear physics continue to attract aspiring physicists. However, you will not be surprised to learn that exploring the galaxies and looking for the basic building-blocks of matter are activities employing very few physicists outside universities. A small number work at the Engineering and Physical Sciences Research Council's Rutherford Appleton Laboratory. A few astronomers are employed by the Astronomy Technology Centre in Edinburgh, and university departments, but many would-be astronomers eventually use their expertise in areas that were subsidiary to their main studies, such as computer programming or electronics.

Government laboratories, such as the National Physical Laboratory and the Defence Science and Technology Laboratory (www.dstl.gov.uk), are active recruiters of physicists, as is QinetiQ (www.qinetiq.com), formerly a government agency but now an independent company. A number of private sector research companies are engaged in contract research for their clients, which frequently requires the expertise of physicists.

Much of what research physicists do in their early careers is laboratory based: applying theories to practical situations, developing novel technologies, troubleshooting and carrying out experiments. As they progress they are more likely to be involved in managing research projects rather than carrying out the day-to-day research themselves.

Patent work and writing Patent work, technical writing and technical journalism can be attractive for those with a talent for communication. Rather than experimental skills, you will need to have sound technical knowledge combined with the ability to write clearly, grammatically and unambiguously. Aspiring patent agents must be prepared to study for the highly competitive professional examinations of the Chartered Institute of Patent Agents and to become a European Patent Attorney. The CIPA have a website at www.cipa.org.uk.

Monitoring health and safety at work The Health and Safety Executive recruits physicists with broad scientific experience to ensure a safe and healthy working environment. The work involves visiting employers, inspecting their premises and investigating the causes of accidents. Specific areas of responsibility include the safety of nuclear installations and North Sea oil and gas platforms. The work sometimes includes taking action through the courts.

Computing Many physicists are interested in careers in computing and information technology. However, demand in these areas can be rather volatile and the number of jobs available varies from year to year. Some physicists go into the computing industry itself, working for employers such as software houses and consultancies and computer manufacturers. Others are employed by large organisations running their own computer departments. There are also opportunities with electronics companies and manufacturers of telecommunications equipment. Working as programmers or systems analysts, physicists may not be using their physics directly, but they are solving problems, applying their logical, numerate approach and using many of the skills they learned as undergraduates.

Finance and other commercial careers Around 40% of all graduate vacancies are open to graduates regardless of subject background. The numeracy and analytical abilities combined with the broad range of other skills offered by their degree courses can make physics graduates particularly attractive to employers in the finance area, such as banks, building societies, insurance companies and accountancy firms.

Some physics graduates enter commercial careers in areas such as retailing, marketing, buying and sales, and can be particularly in demand where the product has a high technical content, such as in the sale of medical diagnostic equipment, sensors and control systems, electronic devices and computers.

There is thus a wide range of career options for physicists with the right mixture of technical and intellectual abilities and personal skills, either using their subject or branching out into new areas. Those who explore their career options early and who gain a variety of work experience whilst at university are best placed to secure a satisfying job on completion of their degree.

Is physics for you? Only you can answer this, but if you are interested in the way things work and enjoy solving puzzles, there is a good chance that it is. Imagination and creativity are a great help, and a willingness to challenge and throw away your own preconceptions is essential. The subject is vast and diverse. At the applied end of the subject it is close to engineering; at the theoretical end it is close to mathematics. It is a good preparation for many careers in technical areas, but also teaches you a variety of skills such as a high level of numeracy and computer literacy that can be used in many other fields. You will also learn about theories that represent as fine a testament to the human intellect and spirit as any painting or piece of music. Unfortunately, whereas paintings and music are to a large extent accessible to everyone, only those who take a degree course in physics can fully appreciate the beauty of these theories.

Chapter 2: The courses

TABLE 2a lists the courses at universities and colleges in the UK that lead to the award of an honours degree in which you spend at least half of your time studying physics. When the table was compiled it was as up to date as possible, but sometimes new courses are announced and existing courses withdrawn, so before you finally decide which courses to apply for you should check the UCAS website, www.ucas.com, to make sure they are still on offer.

Duration The *Duration* column shows the duration of the course, including any time spent abroad or in employment (during a sandwich course), so courses with optional periods of this type will generally show two possible durations, for example '3, 4'. This indicates that a course can, for example:

- take three years' continuous study in the UK, or four years if the option for study abroad is taken
- be taken as a sandwich course or a straightforward full-time degree
- lead to alternative final degrees.

Sometimes several variants of a course are available, differing in length as well as in other features, so you should check with the institution to make sure you know exactly what is on offer.

Foundation years and franchised courses The *Foundation year* column shows whether an optional foundation year is available if your qualifications are not in the relevant subjects for direct entry to the course. A foundation year allows you to acquire the necessary knowledge and skills to begin the main course on a par with students entering the course directly. Foundation years have been introduced mainly to encourage more students to follow science, engineering and mathematics, so are available mostly for those subjects.

Foundation years are not the same as Foundation Degrees. These are qualifications in their own right and are currently only offered in England, Wales and Northern Ireland. Courses leading to a Foundation Degree last two years and are particularly related to workplace skills. However, they can provide another route into honours degree courses, especially at the institutions offering them. For more information, see www.foundationdegree.org.uk.

Franchised courses TABLE 2a also shows whether the foundation year can or must be taken at a franchised institution, which will typically be a college of further education in the same region. Some institutions also franchise complete degree courses, sometimes at colleges outside their immediate locality. Refer to prospectuses for further details of franchising arrangements.

Optional foundation years are not included in the *Duration* column, so you should add one to those figures to give the length including a foundation year.

Direct entry to year 2 in Scotland In Scotland, most students enter university with a broader and less specialised background than in the rest of the UK, so the first year is often similar in function to a foundation course. If you are well qualified, you may be able to gain exemption from this first year for some courses, particularly at the older Scottish universities. The *Foundation year* column shows courses for which this is possible, and in these cases the *Duration* figures should be reduced by one for direct entry to the second year.

Modes of study The *Modes of study* column shows whether a course is available in full-time, part-time, sandwich or time abroad modes. The Guide does not include courses that are only available part time, but does show if courses are available part time in addition to one of the other modes.

Courses may be shown as involving time abroad for a number of reasons. The most straightforward is if the named course has an optional or compulsory period of study or work experience overseas. The period spent abroad will usually be about a year; courses in which you spend less than six months abroad do not qualify for the time abroad description. However, to avoid duplication of information, a time abroad entry may also mean that there is a variant of the course with a slightly different title (usually including a phrase such as 'with study in Europe' or 'International') and involving study or a work placement abroad. Note that a course is shown as including time abroad only when that opportunity arises *as part of the physics course.* If time can be spent abroad only if you combine the study of physics with a modern language, the course is not shown as a time abroad course in TABLE 2a.

Modular schemes Many institutions offer modular schemes in which physics can be studied alongside a wide range of other, often quite unrelated, subjects. Note that although many courses are described as 'modular', this usually means that the physics content itself is organised in modules; a modular *course*, in this sense, may or may not be part of a modular *scheme.*

Modular schemes can give you much greater flexibility in choosing what you study and when, and have the particular benefit of allowing you to delay specialisation until you know more about the subject. However, there is a great deal of variation in the way modular schemes are organised, and there is always some restriction on the combination of modules you can take. If you feel strongly that you want to take a particular combination of subjects in a modular scheme, you should check prospectuses carefully, as some institutions do not guarantee that all advertised combinations will be available.

Course type This Guide describes courses in which you can spend at least half of your time studying physics. Specialised courses, in which you can spend substantially more than half of your time studying one of these subjects, are shown as ● in TABLE 2a. The *No of combined courses* column in TABLE 2a shows how many subjects can be combined with physics in a combined course. In these you spend between half and two-thirds of your time studying physics, and the rest studying another subject. You can use TABLE 2b to find out where you can study a specific subject in combination with physics (you should use the UCAS website or prospectuses if you want to know what combinations are available at a specific institution).

Some entries in TABLE 2a show that you can study physics as a single-subject specialised course or in combination with other subjects. In general, information in later tables for these entries will be given for the specialised course, though in many cases it will apply equally to the combined courses.

Table 2a First-degree courses in **Physics**

Institution Course title (① ② ③ see combined subject list – Table 2b)	**Degree**	**Duration** Number of years	**Foundation year** ● at this institution	○ at franchised institution	◑ second-year entry	**Modes of study** ● full-time	◒ part-time	○ time abroad	◐ sandwich	⊛ **Modular scheme**	**Course type** ● specialised	◐ combined	**No of combined courses**
Aberdeen *www.abdn.ac.uk*													
Natural philosophy	MA	4			◑	●	◒	○		⊛	●		0
Physics ①	BSc/MA	4			◑	●	◒	○		⊛	●	◐	9
Aberystwyth *www.aber.ac.uk*													
Physics ①	BSc/MPhys	3, 4	●			●		○			●	◐	8
Physics with planetary and space physics	BSc/MPhys	3, 4	●			●					●		0
Bath *www.bath.ac.uk*													
Physics	BSc/MPhys	3, 4		○		●		○	◐	⊛	●	◐	2
Belfast *www.qub.ac.uk*													
Physics ①	BSc/MSci	3, 4	●			●		○			●	◐	2
Physics with astrophysics	BSc/MSci	3, 4	●			●					●		0
Theoretical physics	BSc/MSci	3, 4	●			●					●		0
Birmingham *www.bham.ac.uk*													
Physics ①	BSc/MSci	3, 4	●			●		○			●	◐	1
Physics and astrophysics	BSc/MSci	3, 4	●			●					●		0
Physics and space research	BSc/MSci	3, 4	●			●					●		0
Physics with nanotechnology	BSc/MSci	3, 4	●			●					●		0
Physics with particle physics and cosmology	BSc/MSci	3, 4	●			●					●		0
Theoretical physics	BSc/MSci	3, 4	●			●		○			●		0
Theoretical physics and applied mathematics ②	BSc/MSci	3, 4	●			●						◐	1
Bristol *www.bris.ac.uk*													
Physics ①	BSc/MSci	3, 4	●	○		●		○			●	◐	2
Physics with astrophysics	BSc/MSci	3, 4	●	○		●			◐		●		0
Cambridge *www.cam.ac.uk*													
Natural sciences (astrophysics)	BA/MSci	3, 4				●				⊛	●		0
Natural sciences (physics, experimental and theoretical)	BA/MSci	3, 4				●				⊛	●		0
Cardiff *www.cardiff.ac.uk*													
Astrophysics	BSc/MPhys	3, 4	●			●					●		0
Physics ①	BSc/MPhys	3, 4	●			●					●	◐	4
Physics with astronomy ②	BSc/MPhys	3, 4	●			●					●	◐	1
Physics with medical physics ③	BSc	3	●			●					●	◐	1
Theoretical and computational physics	BSc	3	●			●					●		0
Central Lancashire *www.uclan.ac.uk*													
Applied physics	BSc/MPhys	3, 4	●			●	◒				●		0
Astrophysics	BSc/MPhys	3, 4	●			●	◒				●		0
Mathematics and astronomy ①	BSc	3, 4	●	○		●	◒					◐	1
Physics ②	BSc/MPhys	3, 4	●			●	◒			⊛	●	◐	5
Physics with astrophysics	BSc/MPhys	3, 4	●			●					●		0

(continued) **Table 2a**

First-degree courses in **Physics**

Institution Course title (① ② ③ see combined subject list – Table 2b)	**Degree**	**Duration** Number of years	**Foundation year** ● at this institution	○ at franchised institution	◑ second-year entry	**Modes of study** ● full-time	◗ part-time	○ time abroad	◐ sandwich	✤ **Modular scheme**	**Course type** ● specialised	◐ combined	**No of combined courses**
Dundee *www.dundee.ac.uk*													
Applied physics	BSc	4			◑	●	◗	○		✤	●		0
Physics ①	BSc/MSci	4, 5			◑	●	◗	○		✤	●	◐	3
Durham *www.durham.ac.uk*													
Natural sciences (physics) ①	BSc/MSci	3, 4				●				✤		◐	12
Physics	BSc/MPhys	3, 4				●					●		0
Physics and astronomy	BSc/MPhys	3, 4				●					●		0
Theoretical physics	MPhys	4				●					●		0
Edinburgh *www.ed.ac.uk*													
Astrophysics	BSc/MPhys	4, 5			◑	●		○			●		0
Computational physics	BSc/MPhys	4, 5			◑	●		○			●		0
Mathematical physics	BSc/MPhys	4, 5			◑	●		○			●		0
Physics ①	BSc/MPhys	4, 5			◑	●		○			●	◐	4
Exeter *www.exeter.ac.uk*													
Physics ①	BSc/MPhys	3, 4				●		○	◐		●	◐	1
Physics with astrophysics	BSc/MPhys	3, 4				●					●		0
Physics with medical applications	BSc	3				●					●		0
Physics with medical physics	MPhys	4				●					●		0
Physics with quantum and laser technology	BSc	3				●					●		0
Quantum science and lasers	MPhys	4				●					●		0
Glamorgan *www.glam.ac.uk*													
Astronomy	BSc	3	●			●						◐	4
Glasgow *www.gla.ac.uk*													
Astronomy ①	BSc/MA/MSci	4			◑	●						◐	3
Physics ②	BSc/MSci	4			◑	●					●	◐	14
Physics with astrophysics	BSc	4				●					●		0
Heriot-Watt *www.hw.ac.uk*													
Computational physics	BSc/MPhys	4, 5			◑	●		○			●		0
Engineering physics	BSc/MPhys	4, 5			◑	●		○			●		0
Mathematical physics	BSc/MPhys	4, 5			◑	●		○			●		0
Nanoscience	BSc/MPhys	4, 5			◑	●		○			●		0
Photonics and lasers	BSc/MPhys	4, 5			◑	●		○			●		0
Physics ①	BSc/MPhys	4, 5			◑	●		○			●	◐	2
Hertfordshire *www.herts.ac.uk*													
Astronomy ①	BSc	3, 4		○		●	◗	○	◐	✤	●	◐	14
Astrophysics ②	BSc/MPhys	3, 4		○		●	◗	○	◐		●	◐	1
Physics ③	BSc/MPhys	3, 4				●		○	◐		●	◐	14
Hull *www.hull.ac.uk*													
Applied physics	BSc/MPhys	3, 4		○		●		○			●		0
Physics	BSc/MPhys	3, 4		○		●		○			●		0
Physics with astrophysics	BSc/MPhys	3, 4		○		●		○			●		0
Physics with lasers and photonics	BSc/MPhys	3, 4		○		●		○			●		0
Physics with medical technology	BSc/MPhys	3, 4		○		●		○			●		0
Physics with nanotechnology	BSc/MPhys	3, 4		○		●		○			●		0
Imperial College London *www.imperial.ac.uk*													
Physics	BSc/MSci	3, 4				●		○			●		0
Physics with studies in musical performance	BSc	4				●					●		0
Physics with theoretical physics	BSc/MSci	3, 4				●					●		0

(continued) **Table 2a**

First-degree courses in **Physics**

Institution Course title (① ② ③ see combined subject list – Table 2b)	**Degree**	**Duration** Number of years	**Foundation year** ● at this institution	○ at franchised institution	◐ second-year entry	**Modes of study** ● full-time	part-time	○ time abroad	◐ sandwich	**Modular scheme**	**Course type** ● specialised	◐ combined	**No of combined courses**
Keele *www.keele.ac.uk*													
Astrophysics ①	BSc/MPhys	3, 4	●			●		○		●	●	◐	27
Physics ②	BSc/MPhys	3, 4	●			●		○		●	●	◐	27
Kent *www.ukc.ac.uk*													
Astronomy, space science and astrophysics	BSc/MPhys	3, 4	●			●		○			●		0
Physics ①	BSc/MPhys	3, 4	●			●		○			●	◐	1
Physics with astrophysics	BSc/MPhys	3, 4	●			●		○			●		0
Physics with space science and systems	BSc/MPhys	3, 4	●			●		○			●		0
King's College London *www.kcl.ac.uk*													
Mathematics and physics with astrophysics	BSc	3				●					●		0
Physics ①	BSc/MSci	3, 4				●		○			●	◐	4
Physics with astrophysics	BSc	3				●					●		0
Physics with medical applications	BSc	3				●					●		0
Physics with molecular biophysics	BSc	3				●					●		0
Lancaster *www.lancs.ac.uk*													
Physics	BSc/MPhys	3, 4				●		○			●		0
Physics studies	BSc	3				●					●		0
Physics with medical physics	BSc/MPhys	3, 4				●					●		0
Physics with particle physics and cosmology	BSc/MPhys	3, 4				●					●		0
Physics, astrophysics and cosmology	BSc/MPhys	3, 4				●					●		0
Theoretical physics	BSc/MSci	3, 4				●					●		0
Theoretical physics with mathematics	MSci	4				●					●		0
Leeds *www.leeds.ac.uk*													
Nanotechnology	BSc	3	●	○		●					●		0
Physics ①	BSc/MPhys	3, 4	●	○		●		○		●	●	◐	8
Physics with astrophysics	BSc/MPhys	3, 4	●	○		●		○		●	●		0
Theoretical physics	MPhys	4	●	○		●		○	◐	●	●		0
Leicester *www.le.ac.uk*													
Physics	BSc/MPhys	3, 4		○		●		○			●		0
Physics with astrophysics	BSc/MPhys	3, 4		○		●		○			●		0
Physics with nanoscience and technology	BSc/MPhys	3, 4				●		○			●		0
Physics with planetary science	BSc/MPhys	3, 4		○		●		○			●		0
Physics with space science and technology	BSc/MPhys	3, 4		○		●		○			●		0
Liverpool *www.liv.ac.uk*													
Astrophysics	MPhys	4				●					●		0
Mathematical physics	BSc/MMath	3, 4				●					●		0
Physics ①	BSc/MPhys	3, 4		○		●		○			●	◐	2
Physics for new technology	BSc	3		○		●					●		0
Physics with astronomy	BSc	3				●					●		0
Physics with medical applications	BSc	3				●					●		0
Theoretical physics	MPhys	4				●					●		0
Liverpool John Moores *www.ljmu.ac.uk*													
Astrophysics	MPhys	4				●				●	●		0
Physics with astronomy	BSc	3				●				●	●		0
Loughborough *www.lboro.ac.uk*													
Engineering physics	BSc/MPhys	3, 4, 5	●			●		○	◐	●	●		0
Physics ①	BSc/MPhys	3, 4, 5	●			●		○	◐		●	◐	5
Manchester *www.man.ac.uk*													
Physics ①	BSc/MPhys	3, 4	●			●		○		●	●	◐	3

(continued) Table 2a

First-degree courses in **Physics**

Institution Course title (① ② ③ see combined subject list – Table 2b)	**Degree**	**Duration** Number of years	**Foundation year** ● at this institution	○ at franchised institution	◑ second-year entry	**Modes of study** ● full-time	◒ part-time	○ time abroad	◐ sandwich	⊛ **Modular scheme**	**Course type** ● specialised	◐ combined	**No of combined courses**
Manchester (continued)													
Physics with astrophysics	BSc/MPhys	3, 4	●			●		○		⊛	●		0
Physics with technological physics	BSc/MPhys	3, 4	●			●		○		⊛	●		0
Physics with theoretical physics	BSc/MPhys	3, 4	●			●		○		⊛	●		0
Nottingham *www.nottingham.ac.uk*													
Chemistry and molecular physics ①	BSc/MSci	3, 4				●				⊛		◐	1
Mathematical physics	BSc/MSci	3, 4				●					●		0
Physics ②	BSc/MSci	3, 4				●	◒	○		⊛	●	◐	2
Physics with astronomy	BSc/MSci	3, 4				●	◒				●		0
Physics with medical physics	BSc/MSci	3, 4				●	◒				●		0
Physics with theoretical astrophysics	BSc/MSci	3, 4				●	◒				●		0
Physics with theoretical physics	BSc/MSci	3, 4				●	◒				●		0
Nottingham Trent *www.ntu.ac.uk*													
Astronomy ①	BSc	3, 4				●			◐			◐	4
Physics ②	BSc/MSci	3, 4	●	○		●	◒	○	◐		●	◐	1
Physics with astrophysics	BSc	3, 4				●			◐		●		0
Quantum and cosmological physics	BSc/MSci	3, 4				●			◐		●		0
Technological physics	BSc	3				●					●		0
Oxford *www.ox.ac.uk*													
Physics	BA/MPhys	3, 4				●					●	◐	1
Paisley *www.paisley.ac.uk*													
Physics ①	BSc	4, 5			◑	●			◐	⊛	●	◐	1
Physics with medical technology	BSc	4, 5				●			◐		●		0
Technological physics	MSci	5							◐		●		0
Queen Mary *www.qmul.ac.uk*													
Astronomy	BSc/MSci	3, 4	●			●		○			●		0
Astrophysics ①	BSc/MSci	3, 4	●			●		○		⊛	●	◐	1
Physics ②	BSc/MSci	3, 4	●			●		○		⊛	●	◐	9
Theoretical physics	BSc/MSci	3, 4	●			●		○		⊛	●		0
Reading *www.rdg.ac.uk*													
Physics ①	BSc/MPhys	3, 4	●			●		○			●	◐	3
Physics and the universe	BSc/MPhys	3, 4	●			●					●		0
Theoretical physics	BSc/MPhys	3, 4	●			●					●		0
Royal Holloway *www.rhul.ac.uk*													
Applied physics	MSci	4		○		●					●		0
Astrophysics ①	BSc/MSci	3, 4		○		●					●	◐	1
Physics ②	BSc/MSci	3, 4		○		●		○			●	◐	4
Physics with particle physics	BSc	3				●					●		0
Physics with quantum informatics	BSc	3				●					●		0
Theoretical physics	BSc/MSci	3, 4		○		●					●		0
St Andrews *www.st-and.ac.uk*													
Astrophysics	BSc/MPhys	4, 5			◑	●		○		⊛	●		0
Physics ①	BSc/MPhys	4, 5			◑	●		○		⊛	●	◐	7
Physics with photonics	MPhys	5			◑	●		○		⊛	●		0
Theoretical physics ②	MPhys	5			◑	●		○		⊛	●	◐	1
Salford *www.salford.ac.uk*													
Physics ①	BSc/MPhys	3, 4, 5	●			●	◒	○	◐		●	◐	1
Physics with acoustics	BSc/MPhys	3, 4, 5	●			●	◒	○	◐		●		0
Physics with aviation studies	BSc	3				●					●		0

Physics

(continued) **Table 2a** First-degree courses in **Physics**

Institution Course title (① ② ③ see combined subject list – Table 2b)	**Degree**	**Duration** Number of years	**Foundation year** ● at this institution	○ at franchised institution	◑ second-year entry	**Modes of study** ● full-time	◒ part-time	○ time abroad	◑ sandwich	**Modular scheme**	**Course type** ● specialised	◑ combined	**No of combined courses**
Salford (continued)													
Physics with pilot studies	BSc	3				●					●		0
Physics with space technology	BSc/MPhys	3, 4, 5	●			●	◒	○	◑		●		0
Pure and applied physics	BSc/MPhys	3, 4	●			●					●		0
Sheffield *www.sheffield.ac.uk*													
Astronomy ①	BSc	3	●	○		●				●		◑	2
Physics ②	BSc/MPhys	3, 4	●	○		●		○		●	●	◑	4
Physics and astrophysics	BSc/MPhys	3, 4				●					●		0
Physics with medical physics	BSc/MPhys	3, 4	●	○		●					●		0
Theoretical physics	BSc/MPhys	3, 4	●	○		●		○			●		0
Southampton *www.soton.ac.uk*													
Physics ①	BSc/MPhys	3, 4	●			●	◒				●	◑	1
Physics with astronomy	BSc/MPhys	3, 4	●			●	◒	○			●		0
Physics with nanoscience	MPhys	4				●					●		0
Physics with photonics	BSc/MPhys	3, 4	●			●	◒				●		0
Physics with space science	BSc/MPhys	3, 4	●			●	◒				●		0
Strathclyde *www.strath.ac.uk*													
Applied physics	BSc/MSci	4, 5		○	◑	●		○			●		0
Biophysics	MSci	5				●		○			●		0
Laser physics and optoelectronics	BSc	4		○	◑	●		○			●		0
Photonics	MSci	5			◑	●		○			●		0
Physics ①	BSc/MSci	4, 5		○	◑	●		○			●	◑	3
Physics with visual simulation	MSci	5				●		○			●		0
Surrey *www.surrey.ac.uk*													
Physics ①	BSc/MPhys	3, 4	●			●		○	◑		●	◑	1
Physics with medical physics	BSc/MPhys	3, 4	●			●		○	◑		●		0
Physics with nuclear astrophysics	BSc/MPhys	3, 4	●			●		○	◑		●		0
Physics with satellite technology	BSc/MPhys	3, 4	●			●		○	◑		●		0
Sussex *www.sussex.ac.uk*													
Astrophysics	BSc/MPhys/MMath	4				●			◑		●		0
Physics ①	BSc/MPhys	3, 4	●			●		○	◑		●	◑	2
Physics with astrophysics	BSc/MPhys	3, 4				●					●		0
Theoretical physics	BSc/MPhys	3, 4	●			●					●		0
Swansea *www.swan.ac.uk*													
Physics ①	BSc/MPhys	3, 4	●			●		○	◑		●	◑	3
Physics with nanotechnology	BSc	3				●					●		0
Physics with particle physics and cosmology	BSc	3	●			●					●		0
Theoretical physics	BSc/MPhys	3, 4				●					●		0
UCL *www.ucl.ac.uk*													
Astronomy ①	BSc/MSci	3, 4				●					●	◑	1
Astrophysics ②	BSc/MSci	3, 4				●					●	◑	1
Medical physics	MSci	4				●					●		0
Physics ③	BSc/MSci	3, 4				●	◒	○			●	◑	2
Physics with medical physics	BSc	3				●					●		0
Theoretical physics	BSc/MSci	3, 4				●					●		0
Warwick *www.warwick.ac.uk*													
Physics	BSc/MPhys	3, 4				●		○			●	◑	2
York *www.york.ac.uk*													
Physics ①	BSc/MPhys	3, 4	●			●		○		●	●	◑	3

(continued) **Table 2a** First-degree courses in **Physics**

Institution Course title	① ② ③ see combined subject list – Table 2b	**Degree**	**Duration** Number of years	**Foundation year** ● at this institution ○ at franchised institution ◑ second-year entry	**Modes of study** ● full-time; ◗ part-time ○ time abroad ◐ sandwich	◈ **Modular scheme**	**Course type** ● specialised; ◐ combined	**No of combined courses**
York (continued)								
Physics with astrophysics		BSc/MPhys	3, 4	●	● ○	◈	●	0
Theoretical and computational physics		BSc/MPhys	3, 4	●	● ○	◈	●	0

Subjects available in combination with physics TABLE 2b shows subjects that can make up between a third and half of your programme of study in combination with physics in the combined degrees listed in TABLE 2a. For example, if you are interested in combining physics with geography, first look up geography in the list to find the institutions offering geography in combination with physics. You can then use the index number given after the institution name to find which of the courses at the institution can be combined with geography. If there is no index number after an institution's name in TABLE 2b, that is because there is only one course at that institution in the appropriate part of TABLE 2a.

It is not possible to describe in the space available here the many different ways in which combined courses are organised, so you should read prospectuses carefully. For example, you should find out if the subjects are taught independently of each other or if they are integrated in any way. Combined courses in modular schemes often provide considerable flexibility, allowing you to vary the proportions of the subjects and include elements of other subjects. However, this means that you may lose some of the benefits of more integrated courses.

Note that the names given in the table for the combined subjects have been standardised to make comparison and selection easier. This means that the name used at a particular institution may not be exactly the same as that given in the table. However, in nearly all cases it will be very similar, so you should not find much difficulty in identifying a particular course combination when you look at the prospectus.

Table 2b Subjects to combine with **Physics**

American studies Keele① ②
Anthropology Durham①, Glasgow②
Applied mathematics Belfast①, Birmingham②, Glasgow① ②
Archaeology Durham①
Artificial intelligence Leeds①
Astronomy Cardiff②, Glasgow②, Sussex①, UCL③
Biochemistry Keele① ②
Biology Durham①, Keele① ②
Business computing Keele②
Business studies Aberystwyth①, Birmingham①, Durham①, Glasgow②, Hertfordshire① ③, Keele① ②, Manchester①, Queen Mary②, Warwick, York①
Chemistry Aberdeen①, Cardiff①, Durham①, Keele① ②, Leeds①, Nottingham①, Nottingham Trent①, St Andrews①
Computer science Aberystwyth①, Belfast①, Durham①, Edinburgh①, Glasgow②, Keele① ②, King's College London①, Leeds①, Queen Mary②, Reading①, Royal Holloway②, St Andrews①, Sheffield②, Swansea①
Computing Bath, Cardiff①, Hertfordshire① ② ③, Loughborough①, Nottingham Trent①, Queen Mary②

(continued) **Table 2b**

Subjects to combine with **Physics**

Criminology Keele① ②
Cultural studies Glamorgan
Earth science Durham①
Economics Durham①, Hertfordshire① ③, Keele① ②, Queen Mary②
Education Aberdeen①, Aberystwyth①, Heriot-Watt①, Keele① ②, Strathclyde①
Electronic engineering Dundee①
Electronic music Hertfordshire① ③
Electronics Queen Mary②
English Keele① ②
English language Hertfordshire① ③
Enterprise/entrepreneurship Sheffield②
Environmental science Heriot-Watt①, Keele① ②
Environmental studies Hertfordshire① ③, Queen Mary②
European languages Nottingham②
Film/television studies Glasgow②
Finance Queen Mary②, Strathclyde①, Surrey①
Forensic science Kent①, Nottingham Trent②, Sussex①
French Aberdeen①, Aberystwyth①, Keele① ②, Leeds①, St Andrews①
Gaelic Aberdeen①
Geography Durham①
Geology Aberdeen①, Glamorgan, Keele① ②, Royal Holloway①
German Aberdeen①, Aberystwyth①, Leeds①
History Keele① ②
History/philosophy of science Leeds①, St Andrews①
Human biology Hertfordshire① ③
Human resource management Keele① ②
Information systems Keele①, Loughborough①
Internet technology St Andrews①
Law Hertfordshire① ③
Management science Hertfordshire① ③, Keele① ②
Management studies Central Lancashire②, King's College London①, Loughborough①, Royal Holloway②
Manufacturing studies Hertfordshire① ③
Marketing Central Lancashire②, Keele① ②
Materials science Queen Mary②
Mathematics Aberdeen①, Aberystwyth①, Bath, Bristol①, Cardiff①, Central Lancashire① ②, Dundee①, Durham①, Edinburgh①, Exeter①, Glasgow① ②, Hertfordshire① ③, Keele① ②, King's College London①, Leeds①, Liverpool①, Loughborough①, Manchester①, Nottingham Trent①, Queen Mary① ②, Reading①, Royal Holloway②, St Andrews① ②, Sheffield① ②, Southampton①, Strathclyde①, Swansea①, UCL② ③, Warwick, York①
Media studies Keele① ②
Medical physics Cardiff③
Medicinal chemistry Keele① ②
Meteorology Edinburgh①, Reading①
Microelectronics Dundee①
Modern languages Salford①
Multimedia Paisley①
Music Cardiff①, Edinburgh①, Glasgow②, Keele① ②, Royal Holloway②
Music technology Keele① ②
Neuroscience Keele① ②
Oceanography Liverpool①
Philosophy Aberdeen①, Bristol①, Central Lancashire②, Durham①, Keele① ②, King's College London①, Leeds①, Manchester①, Nottingham②, Oxford, Sheffield① ②, York①
Physical geography Aberystwyth①, Glamorgan
Physics Glasgow①, Nottingham Trent①, UCL①
Polish Glasgow②
Politics Keele① ②
Psychology Durham①, Glasgow②, Hertfordshire① ③
Scottish history Glasgow②
Sociology Keele① ②
Space physics Glamorgan
Spanish Aberdeen①, Aberystwyth①, St Andrews①
Sports science Loughborough①, Swansea①
Sports studies Hertfordshire① ③
Theatre studies Glasgow②
Theology Glasgow②
Tourism Hertfordshire① ③

Other courses that may interest you The following Guides contain courses that may interest you as an alternative to physics:

- *Engineering:* the Engineering Guides include courses in electronics as well as all branches of engineering
- *Biological Sciences:* for biophysics
- *Geography and Geological Sciences:* for geophysics
- *Mathematics, Statistics and Computer Science:* for courses in applied mathematics.

Physics has well-established and widely agreed foundations, so the basic content of courses changes only slowly, and varies little from one institution to another. In contrast, in the final year or two of the courses there is usually a wide range of specialist topics that take you to the frontiers of knowledge. These topics often reflect the research interests of individual departments and members of staff, and may change from year to year. The flexibility provided by modular programmes and the extra time given by MSci/MPhys four-year degrees has encouraged the development of a wider variety of these specialised topics.

The next few sections describe the different types of course covered by this Guide and the basic content they include. TABLE 3a gives information about the final-year content of courses. Following this, there are sections on MSci/MPhys courses, on work experience, including sandwich courses, and on the possibilities of spending time abroad.

The range of courses This Guide covers courses in the general area of physics, including applied physics, chemical physics, astronomy, astrophysics, mathematical and theoretical physics, and physics with electronics, as well as courses in pure physics. There is considerable overlap in the content between these types of course, and at institutions with more than one type of course, groups of students are often brought together for the teaching of some topics. The following section describes the structure and content of typical physics courses. Subsequent sections explain the differences in emphasis in the more specialised types of course.

Physics The first year Courses begin by building on the foundations of A-level physics. You may find yourself going over some familiar material, as institutions have to allow for the fact that students will have taken different A-level courses, but the pace at which it is covered will be much faster, so it will not be long before you begin to break new ground.

Typically, first-year courses will include work on properties of matter, wave theory and optics, electromagnetism and classical mechanics (the word 'classical' is used to distinguish the subject from quantum mechanics – classical mechanics is not restricted to Newtonian mechanics, so courses may include some work on special relativity). Many places introduce some elementary atomic physics, quantum mechanics and relativity in the first year.

Laboratory work will probably be related to theory work, but much greater emphasis than you will have been used to at school will be given to a critical analysis of errors, both to try to reduce them and to estimate those that remain, so that you can state quantitatively how significant your results are. To be able to do this you will need to be taught the relevant parts of probability theory and statistics. You will use sophisticated equipment from the start of the course, such as computer interfacing techniques for direct data handling.

The work in physics will be supported by mathematics lectures in areas such as vectors (for use in mechanics, electricity and magnetism), complex numbers (used in AC theory, wave motion and optics) and partial derivatives (for estimating experimental tolerances, wave motion and thermodynamics). You will also study other topics as a foundation for the more theoretical parts of physics that come later, such as determinants and matrices, groups, sequences and series, advanced calculus, and the solution of ordinary and partial differential equations. Computational methods will also be introduced at some point, though not always in the first year.

The second year   The work in the second year is usually much more advanced. Typical topics include quantum or wave mechanics, thermodynamics and statistical mechanics, special relativity, condensed matter and nuclear physics, as well as more advanced electromagnetism and optics.

There may be further lectures on mathematical methods, particularly in courses with a more theoretical emphasis, and laboratory work will continue in parallel with the theory lectures.

The final year   The final year is generally much more specialised and often has considerable scope for following your own interests through options. TABLE 3a gives more information about final-year topics for all the types of course covered in this Guide. Another way in which you can develop a particular interest is through a project or dissertation, which features as a compulsory or optional element in many courses: Chapter 4 has more information about projects.

By this stage in the course, it is unlikely that there will be separate lectures on mathematical methods, though theory options will have a high mathematical content. Laboratory work is likely to be much more specialised and may be largely or wholly replaced by projects.

Applied physics   The basic content of applied physics courses is usually similar to that of pure physics courses but, as you would expect, stresses applications. Emphasis is given to the more applicable parts of physics such as electronics, solid-state devices, nuclear energy and properties of materials. This is at the expense of the more theoretical areas such as relativity, and the mathematical content may be less. However, it would be wrong to think the courses are any easier; the demands are different but just as challenging.

Applied physics courses are often more broadly based and may include many more lectures from other departments such as chemistry, electronic engineering, instrumentation, engineering drawing, industrial administration, legal practices and social studies.

The courses are designed mainly for those who intend to work in industrial or government research and development laboratories, but also form an excellent basis for a career in teaching. However, the courses are not so specialised that you could not, for example, go on to academic research in a pure physics department.

Many of the courses are sandwich courses or have a sandwich option: more details are given later in this chapter.

There is often considerable overlap of content between applied physics courses and physics with electronics courses.

Mathematical or theoretical physics Some institutions run courses with a more mathematical bias, which can take a variety of forms. Some have theoretical options in the final year, often replacing part or all of the laboratory work. Such courses generally differ from combined physics and mathematics courses, which are usually constructed from topics taught separately to specialised honours degree students in the two departments, with little integration. In contrast, mathematical or theoretical physics courses are typically planned as a coherent programme based on the more theoretical parts of physics and the more applied parts of mathematics.

Topics covered may include quantum theory, general relativity, electromagnetic field theory, quantum electronics, statistical mechanics, fluid mechanics, theoretical nuclear physics, elementary particle physics, solid-state physics and group theory. Another feature of mathematical and theoretical physics courses is the emphasis placed on computational aspects, with extensive use of various software techniques.

If you think you may want to follow a career in theoretical physics, you should make sure the course you intend to follow is either a specialised theoretical course or contains a substantial proportion of theoretical options. Alternatively, you could take a first degree in mathematics rather than physics. If you are thinking of taking a theoretical course (or options), however, you should be aware that the mathematical demands are much greater.

Astronomy and astrophysics Astronomy and astrophysics feature as components of many physics courses, but are also available as specialised courses on their own; as combinations, often with physics or mathematics; or as minors with a physics major. The content varies, but it usually includes studies of observational techniques, cosmology, astronomical spectroscopy, solar and planetary dynamics, radio/X-ray/optical astronomy, stellar structure and evolution, quasars and black holes. If you are interested in the more theoretical aspects, particularly in areas such as cosmology, you should also consider the possibility of taking an applied mathematics or theoretical physics course.

Chemical physics The courses generally constitute an integrated programme bringing together various parts of specialised courses in chemistry and physics. Emphasis is on electromagnetism, classical physics, solid-state physics, thermodynamics, wave mechanics, spectroscopy and the physical chemistry of solids and polymers.

Physics with electronics, electronic physics and related courses Electronics is taught in most physics courses, but if you have a particular interest in this area you can take a course which gives it greater emphasis. One way of doing this is to take an applied physics course, which will usually cover electronics in much greater detail than pure physics courses do. Alternatively, you can look at TABLE 2b for courses where electronics is available as a combined course with physics. You should also consult the *Electrical and Electronic Engineering* section of the *Engineering* Guide for specialist courses in electronics.

Courses that are closely connected with physics give greater emphasis to covering the basic principles of physical electronics than do more engineering-based courses, which emphasise the more practical side of electronics. However, it is only a matter of emphasis, and both types of course will cover all aspects to some extent. In many courses the electronics component is taught entirely within the physics department, but at some institutions some lectures are given by engineering staff. The details of physical electronics courses vary considerably in their specialisation and structure. For example, some courses concentrate specifically on optoelectronics. The trend is for courses to give greater emphasis to digital electronics including microprocessors and programming, but analogue circuits are still covered in detail in most courses. Practical work is an important part of all courses, so a significant part of your time will be spent in the laboratory.

Medical physics Medical physics is a rapidly growing field. It can be studied through a specialised course but is more commonly available as a minor course with major physics or as a specialised option in physics courses: see TABLE 3a. In general, the major part of medical physics courses concentrates on the same fundamental physics found in other physics courses, though there is usually particular emphasis on areas such as nuclear physics, electromagnetic waves and electronics, which have particular relevance for physics-based diagnostic and therapeutic methods, and which are usually treated in the final specialist year.

Final-year specialisation The range of topics available in the final year varies widely from institution to institution, as does the amount of choice you have. TABLE 3a shows what percentage of your time in the final year is spent on compulsory topics in physics, and what percentage can be spent on optional topics (other topics outside physics make up the balance). It also shows which topics are on offer on each course, and whether they are compulsory components ●, optional components ○, or with some compulsory and additional optional components ◑. Options allow you to follow a particular interest to a greater depth and can often add considerably to the enjoyment and benefit you gain from the course.

As mentioned earlier, the topics available in the final year tend to change more frequently than those in earlier years, so you should treat the information in the table as giving a good indication of the general pattern, but you should contact departments directly if a particular topic is important to you. Where a topic is only available on an MSci/MPhys course, it is shown as ◡ in the table. TABLE 3b then summarises the differences between Master's level courses and the corresponding BSc courses.

Table 3a

Course content

Institution
Course title
① ②: see notes after table

○ optional; ● compulsory;
◑ compulsory + options;
◒ only available for MPhys, MSci etc

Institution / Course title	Final year physics content Compulsory (%)	Optional (%)	Experimental physics	Theoretical/mathematical physics	Astronomy/astrophysics	Computational physics	Geophysics	Biophysics	Medical physics	Applied physics	Environmental physics	Solid-state physics	Low temperature physics	Materials	Nanoscience	Atomic physics	Nuclear physics	Elementary particles	Quantum mechanics	Electromagnetism	Relativity	Plasma physics	Optics/lasers	X-ray crystallography	Instrumentation	Acoustics/ultrasonics	Non-linear physics
Aberdeen																											
Natural philosophy ①	62	25	●		●			●	●			●						●		●	○		◑	●	●		
Physics ②	63	37	●	●	●			●	●	●		●		●				●	●	●	○		◑	●	●		
Aberystwyth																											
Physics ①	100		●	●	○	○						●		●		●	●	●	●	●	●	◒	●	●	●		
Physics with planetary and space physics ②	100		●	●	●		●					●				●	●	●	●	●	●	◒	●		●		
Bath																											
Physics ①	10	90	◑	◑	◑	◑			○	○	○	◑	○	○	◒	◑	◑	◑	◑	◑	◑		◑		◑		○
Belfast																											
Physics ①	50	50	●	●	○							○		○		○	○	○	●	●	●	○	○				
Physics with astrophysics ②	64	36	●	●	●							○				○	○	○	●	●	●	○	○				
Theoretical physics ③				●	○							○							●	●	●						
Birmingham																											
Physics ①	58	42	●	○	○		○	○	○	○	○	○	○	○		○	○	○	●	○	○	○	●	○			
Physics and astrophysics ②	52	48	●	○	◑					○	○	○	○			○	○	○	●	●	●	◒	○		◑		
Physics and space research ③	52	48	●	○	◑					○	○	○	○			○	○	○	●	●	●		○		●		
Theoretical physics ④	75	25	○	●	○					○		○	○			○	○	○	●	●	●		○	○			
Theoretical physics and applied mathematics ⑤	50	50		●	○					○		○	○				○	○	●	●	●		○				
Bristol																											
Physics	10	90	○	○	○		○	○	○	○	○	○	○	○		○	○	○	◑	○	○				○		
Physics with astrophysics	10	90	○	○	●		○	○	○	○	○	○	○	○		○	○	●	◑	○	○				○		
Cambridge																											
Both courses ①	50	50	○	○	○		○		○			●	○	○		●	●	●	◑	●	●	●	○				
Cardiff																											
Astrophysics	75	25		●	●												●	●	●	●	●						
Physics	75	25	●	●								●	●				●	●	●	●	●		●				
Physics with astronomy			○	●	●												●	●	●	●	●						
Physics with medical physics	75	25	●	●					●			●	●				●		●	●					●	●	
Theoretical and computational physics				●	○												●	●	●	●	●						
Central Lancashire																											
Applied physics ①	83	17	●							◑		●					●		◒	●					●		
Astrophysics ②	83	17		●	●													●	◒		●	●	●				
Mathematics and astronomy	30	20			●																						
Physics ③	83	17	●							◑		●					●		◒	●							
Dundee																											
Applied physics ①	100		●	●		●				●		●		●					●	●			●		●		
Physics ②	100		●	●		●		◒				●	●	●		●	●		●	●	●		●	●	●		
Durham																											
Natural sciences (physics)	50	50	●						◒	●	○	●		●		○	○						●		◒		
Physics ①	50	50	●	○	◑				◒	○	○	●	○	○		○	○	●	◑	◑	◑		◑	○			
Physics and astronomy ②	50	50	●		●					○	○	●	○	○		○	○	●	◑	◑	◑		◑	○			
Theoretical physics	50	50		●	○					○	○	●	○	○		○	○	○	◑	◑	◑		◑	○			
Edinburgh																											
Astrophysics	65	35	●	○	○		○			○		○		○		●	○	○	○	○	●		○		●	○	
Computational physics ①	35	65	○	○	○	●	○			○		◑		○		●	○	○	○	○	○		○			○	
Mathematical physics ②	45	55		●	○	●	○			○		◑		○		●	○	○	◑	○	○		○			○	
Physics ③	20	80	●	○	○	●	○					◑		○		○	○	○	●	○	○		○			○	

Physics

(continued) Table 3a

Course content

Institution
Course title
① ②: see notes after table

○ optional; ● compulsory;
◑ compulsory + options;
◒ only available for MPhys, MSci etc

Institution / Course title	Final year physics content: Compulsory (%)	Optional (%)	Experimental physics	Theoretical/mathematical physics	Astronomy/astrophysics	Computational physics	Geophysics	Biophysics	Medical physics	Applied physics	Environmental physics	Solid-state physics	Low temperature physics	Materials	Nanoscience	Atomic physics	Nuclear physics	Elementary particles	Quantum mechanics	Electromagnetism	Relativity	Plasma physics	Optics/lasers	X-ray crystallography	Instrumentation	Acoustics/ultrasonics	Non-linear physics
Exeter																											
Physics ①	66	33	●		○			○	○	○	○	●	○			●	●	○	●	●	○		○		○	○	
Physics with astrophysics	75	25	●		●			○	○	○	○	●	○			●	●	○	●	●	●		○		●	○	
Physics with medical applications ②	84	16	●		○			●	●	●	○	●		●			●	○	●	●			●			●	
Physics with medical physics ③	75	25	●		○			●	●	●	○	●	○	●		●	●	○	○	●	○		○		●	●	
Physics with quantum and laser technology ④	84	16	●	○					○		○	●		●		●	●	○	●	●	○		●	●			
Quantum science and lasers ⑤	84	16	●	○					○		○	●	○	●		●	●	○	●	●	○		●	●			
Glasgow																											
Astronomy	67	33		●	●				○	○		○	○	○		○	○	○	◑	●	◒	◒	○		●		
Physics	50	50	●	●	○				○	○		●	○	◑		○	◑	◑	◑	●	◑	○	○		○		
Heriot-Watt																											
Computational physics ①	100		●	●	●	●					●	●	●	●		●	●	●	●	●	●		●	●			
Engineering physics ②	100		●	○	○					●		●		●		●	●	●	●	●	●	●	●		●		
Physics	100		●	○	○	●			○	○	●	●					●	●	●	●			●				
Hertfordshire																											
Astronomy	80	20	●		●		○										○	●	○		●	●	◑		○		
Astrophysics	70	30	●		●		○		○			○					○	●	○		●	●	◑	○	○		
Physics ①																	●			●			●				
Hull																											
Applied physics ①	63	37	●		○					◑	○	◑	●	○		●	●	○	●	●	○	◒	○	●	○	○	
Physics ②	75	25	●	●	○							●	○	○		●	●	●	●	●	◑	○	◑	●	○	○	
Physics with lasers and photonics ③	75	25	●	●	○							◑	●			◑	◑	○	◑	●	○	●	◑	○	○	○	
Physics with medical technology ④	75	25	●	●	○			●	●			●	○	○		●	◑	○	◑	●	○	◑	◑	○	●	○	
Imperial College London																											
Physics ①	20	80	◑	◑	○	○		○	○	○	○	●		○	○	◑	●	◑	◑	●	◑	○	○		○		○
Physics with studies in musical performance ②	80	20	●		○			○	○			●		○		◑	●	●	◑	●	◑	○	○		●		
Physics with theoretical physics	30	70	●	●	○	○		○	○		○	○				◑	●	◑	◑	●	◑	○	○		○		○
Keele																											
Astrophysics	50	50	●	◑	◑	○						○	○	○	○	◑	◑	◑	◑	●	●		◑				
Physics	75	25	●	◑	○	○						◑	◑	◑	◑	○	○	○	○	●	○		○	○			
Kent																											
Physics ①	88	12		●	○				○			●				●	●	●	●	●	●		●				
Physics with astrophysics ②	88	12		●	●				○			●				●	●	○	●	●	●		●				
Physics with space science and systems ③	88	12		●	○				○			●				●	●		●	●	●		●				
King's College London																											
Mathematics and physics with astrophysics	50	50	●	●	●							○						○	●		●						
Physics	80	20	●	○	○				○	○		○	○	○	◒	○	○	○	○	○	○	○	○	○	○		
Physics with astrophysics	80	20	●	○	●	●						○				○	○	○	●	●	●						
Physics with medical applications			●	○					●							○	○	○	●	●	●						
Physics with molecular biophysics	80	0	●	●				●	●			●					●		●	●							
Leeds																											
Physics ①	50	50	●	○	○				○	○		●	○				○	○	●		○		○		○		
Physics with astrophysics ②	75	25	●	○	●				○	○		○	○	○			○	○	●	●	○		○		○		
Theoretical physics	40	60		●	○				○			○	○				○	○	○	○	○		○				

(continued) **Table 3a**

Institution Course title ① ② : see notes after table — ○ optional; ● compulsory; ◑ compulsory + options; ◒ only available for MPhys, MSci etc	**Final year physics content** Compulsory (%)	Optional (%)	Experimental physics	Theoretical/mathematical physics	Astronomy/astrophysics	Computational physics	Geophysics	Biophysics	Medical physics	Applied physics	Environmental physics	Solid-state physics	Low temperature physics	Materials	Nanoscience	Atomic physics	Nuclear physics	Elementary particles	Quantum mechanics	Electromagnetism	Relativity	Plasma physics	Optics/lasers	X-ray crystallography	Instrumentation	Acoustics/ultrasonics	Non-linear physics
Leicester																											
Physics ①	33	67	◑	◑	○	◑	○	○	○	◑	○	○		○	○	○	◑	○	◑	○	○	○	○	○	○		
Physics with astrophysics ②	33	67	◑	○	◑	◑	○	○	○	◑	○	○		○	○	○	◑	○	◑	○	○	○	○	○	○		
Physics with nanoscience and technology			◑	○	○	◑	○	○	○	◑	○	○		○	○	○	◑	○	◑	○	○	○	○	○	○		
Physics with planetary science ③	33	67	◑	○	○	◑	○	○	○	◑	○	○		○	○	○	◑	○	◑	○	○	○	○	○	○		
Physics with space science and technology ④	33	67	◑	○	○	◑	○	○	○	◑	○	○		○	○	○	◑	○	◑	○	○	○	○	○	○		
Liverpool																											
Astrophysics	25	50	●	○	●						○	●				●	●	●	●	●	●						
Mathematical physics	50	50	○	●	○						○	○	○			○	○	○	●	○	●		○				
Physics ①	37	37	●	○	○						○	●	○	○		●	○	○	●	●	●				○		
Physics for new technology ②	60	25	●		○					●	○	●												●	●		
Liverpool John Moores																											
Astrophysics	70	30	●	●	●	●				○	○	●	○	○		●	●	●	●	●	●		●		●		
Physics with astronomy	70	30	●	●	●	●				○	○	●	○	○		●	●	○	●	●	●		●		●		
Loughborough																											
Engineering physics ①	17	50	●									○	○	○		○		○	○	○			○				
Physics ②	25	75	●	◒		○						○	○	○		●		○	○	○	◒		○				
Manchester																											
Physics ①	50	50	●	○	○	◑	○	○	○	○	○	◑	○	○		◑	◑	◑	◑	◑	◑	○	◑	○	○		○
Physics with astrophysics ②	70	30	●	○	◑	◑	○	○	○	○	○	◑	○	○		◑	◑	◑	◑	◑	◑	○	◑	○	○		○
Physics with technological physics ③	60	40	●	○	○	◑	○	○	○	○	○	◑	○	○		◑	◑	◑	◑	◑	◑	○	◑	○	○		○
Physics with theoretical physics ④	50	50	●	◑	○	○	○	○	○	○	○	◑	○	○		◑	◑	◑	◑	◑	◑	○	◑	○	○		○
Nottingham																											
Chemistry and molecular physics	25	25	●	◑	○			○	○		○	◑	○			●			◑	○	○		◑	●	◑		
Mathematical physics ①	45	55		●	◑							●	○			◑	○	●	●	●	◑		◑				
Physics ②	60	40	●	○	○			○	○	○	○	◑	○	●		●	●	◑	◑	●					○		
Physics with astronomy ③	85	15	●		●				○		○	●	○			●	●	●	●	●					●		
Physics with medical physics	85	15	●		○				●		○	●	○	◒		●	●	●	●	●					●		
Physics with theoretical astrophysics ④				●	●														●								
Physics with theoretical physics ⑤				●															●	●							
Nottingham Trent																											
Physics ①	50	50	●							◒	●	●	◒	●			●		●	●			●	●	●	●	
Oxford																											
Physics ①	80	20	◑	○	○		●	○	○		○	◑				◑	◑	◑	◑	●	◑		◑				
Paisley																											
Physics	80	20	●				○			○		●	○			●	●	●	●	●			◑	○	◑		
Queen Mary																											
Astronomy	25	75	○	○	◑					○						○	○	○	○	○	○	○	○		○		
Astrophysics	25	75	○	○	◑					○						○	○	○	○	○	○	○	○		○		
Physics	38	62	◑	○	○				○	○		○	○			○	○	○	◑	○	○	○	○		○		
Theoretical physics ①	25	75		◑	○	○						◑	○	○		○	○	○	◑	◒	○	○	○				
Reading																											
Physics	70	30	●	●	◑		◑		○	◑		◑		◑		●	●	●	●	●			◒				
Theoretical physics ①	90	10		●	●							●							●	●			◒				
Royal Holloway																											
Applied physics ①	0	100	○	○	○					●		○	○			○	○	○	○	○	○		○		○		

Physics

(continued) **Table 3a** Course content

Institution
Course title
① ②: see notes after table

○ optional; ● compulsory;
◑ compulsory + options;
◒ only available for MPhys, MSci etc

Final year physics content

Institution / Course title	Compulsory (%)	Optional (%)	Experimental physics	Theoretical/mathematical physics	Astronomy/astrophysics	Computational physics	Geophysics	Biophysics	Medical physics	Applied physics	Environmental physics	Solid-state physics	Low temperature physics	Materials	Nanoscience	Atomic physics	Nuclear physics	Elementary particles	Quantum mechanics	Electromagnetism	Relativity	Plasma physics	Optics/lasers	X-ray crystallography	Instrumentation	Acoustics/ultrasonics	Non-linear physics
Royal Holloway (continued)																											
Astrophysics ②	0	100	○	○	●					○		○				○	○	○	○	○	○		○				
Physics ③	0	100	○	○	○					○	○	○	○			○	○	○	○	○	○		○	○			
Theoretical physics ④	0	100	○	●	○					○		○	○			○	○	○	○	○	○		○				
St Andrews																											
Astrophysics	70	30		○	●	○		○					○			○	○	○	●	●	●		○				
Physics	30	70	●	○	○	○		○				●	◑			●	●	○	●	●	●		◑				
Physics with photonics	75	25	●	○	○	●		○				●	○			●	○	○	●	●	●		●				
Theoretical physics	50	50	○	●	○	●		○				●	○			●	●	●	●	●	●		○				
Salford																											
Physics ①	75	25	●		○				○	○	○	●	○	●		○	●	○	●	●			◑		○	○	
Physics with acoustics	100	0	●									●		●			●		●	●						●	
Physics with space technology ②	100	0	●		●							●		●			●		●	●							
Sheffield																											
Astronomy ①	17	50	●	○	●	○	○				○	●	○	○		●	○	◑	◑	●	◑		○		●		
Physics ②	50	50	●	○	○	◑	○		○	◒		◑	●	○	○	●	●	◑	◑	●	◑		◑		○		
Physics with medical physics ③	58	16	●	○		◑	○		●	◒		◑	●			●	●	◒	●	●	●		◑		●		
Theoretical physics ④	75	25	●	●	○	●	○		○	◒		◑	●		○	●	●	●	●	●	◑		◑				
Southampton																											
Physics	70	30	●	●	○	◑			○	○	○	●		●	◒	●	●	◑	●	●	●	◒	◑				
Physics with astronomy	90	10	●	●	●	◑					○	●		●	○	●	●	◑	●	●	●	◒	●				
Physics with photonics	80	20	●	●		◑			○		○	●		●	○	●	●	◑	●	●	●	◒	●				
Physics with space science	80	20	●	●	○	◑						●		●		●	●	◑	●	●	●	◒	●				
Strathclyde																											
Applied physics	60	40	●	○	○			○	○	●	●	●	●	○		○	○	○	○	●	○	○	○	○	●	●	
Biophysics	45	55	●			●		●	●	○	○	○	○	○			○			○			○	●	●	●	
Laser physics and optoelectronics	70		●	○	○	●		○	○	●	○	●	○	○		●	○	○	●	●	○	○	●	○	●	○	○
Photonics	70	30	◑			●				●		●		●		●			●	●			●		●	●	●
Physics	60	40	●	○	○	●		○	○	○	○	●	○	○		●	●	●	●	●	○	○	○	○	●	○	○
Physics with visual simulation	60	15	●	●	○	●		○	○	●	○	●	○	●		●	●	○	●	●	●	●	●	○	●	○	●
Surrey																											
Physics	33	42	○	○	○			○	○			○		○			○	●	●		●		○			○	
Physics with medical physics	67	8	○	○	○			●	●			○		○			○	●	●		●		○			○	
Physics with nuclear astrophysics	67	8	○	○	●			○	○			○		○			○	●	●		●		○			○	
Physics with satellite technology	67	8	○	○	●			○	○			○		○			○	●	●		●		○			○	
Sussex																											
Physics	30	70	◑	○	○		○				○	●	○			◒	◑	◑	○			○	●		○		
Theoretical physics	40	60		●	○		○				○	●	○			◒	◑	◑	●			○	●		○		
Swansea																											
Physics ①	80	20	●	●	○	●		○	○	○		●		○	○	●	●	●	●	●	●		○		○		●
Physics with nanotechnology	100		●	●	●	●		●		●		●		●	●	●	●	●	●	●	●		●		●		●
Physics with particle physics and cosmology ②	80	20	●	●	●	●				○		●				●	●	●	●	●	●		●				
Theoretical physics	100		●	●	●	●				●		●					●	●	●	●	●						●
UCL																											
Astronomy		100	◑	○	○		○		○	○	○	○	○	○		○	○	○	○	○	○	○	○	○	○		
Astrophysics		100	◑	○	○		○		○	○	○	○	○	○		○	○	○	○	○	○	○	○	○	○		
Medical physics	40	40	◑	○	○		○		●	○	○	○	○	○		○	○	○	○	○	○	○	○	○	○		
Physics		100	◑	○	○		○		○	○	○	○	○	○		○	○	○	○	○	○	○	○	○	○		

(continued) Table 3a Course content

Institution Course title ① ②: see notes after table ○ optional; ● compulsory; ◑ compulsory + options; ◒ only available for MPhys, MSci etc	**Final year physics content** Compulsory (%)	Optional (%)	Experimental physics	Theoretical/mathematical physics	Astronomy/astrophysics	Computational physics	Geophysics	Biophysics	Medical physics	Applied physics	Environmental physics	Solid-state physics	Low temperature physics	Materials	Nanoscience	Atomic physics	Nuclear physics	Elementary particles	Quantum mechanics	Electromagnetism	Relativity	Plasma physics	Optics/lasers	X-ray crystallography	Instrumentation	Acoustics/ultrasonics	Non-linear physics
UCL (continued)																											
Physics with medical physics	50	50	◑	◑	○		○		●	○	○	○	○	○		◑	◑	○	◑	◑	○	○	○	◑	◑		
Warwick																											
Physics	20	80	◑	○	○	○	○		○	○		○	○	○		○	○	○	◑	◑	○	○	○	○		○	
York																											
Physics ①	66	34	●	◑	○		○	○			○	◑				◑	◑	●	◑	◑	○	◒	◑	○	○		
Physics with astrophysics	100		●	●	●							●				●	●	●	●	●	●		●				
Theoretical and computational physics ②	66	34		◑	○			○				◑				◑	◑	●	◑	◑	○	◒	◑	○			

Aberdeen ① History and philosophy of science ● ② Case studies ●

Aberystwyth ① ② Problem solving ●; fluid mechanics ◒

Bath ① Statistical mechanics ●; fluid dynamics ○; management, economics and marketing ○; electron beam physics ◒; acoustic scattering ◒; non-linear physics ◒; quantum nanostructure devices ◒; surface and interface physics ◒

Belfast ① Signal processing ○; optoelectronics ○; physical electronics ○ ② Signal processing ○; optoelectronics ○ ③ Calculus of variations ●; numerical analysis ○; dynamical systems ○; statistical mechanics ◒

Birmingham ① 7-week group research activity ●; individual research project ◒ ② Observatory laboratory; 7-week group research activity; individual research project ◒ ③ Space research group design activity (7 weeks); individual research project ◒ ④ ⑤ Research project ◒

Cambridge ① Statistical physics ●

Central Lancashire ① Statistical mechanics ◒; virtual instrumentation ● ② Statistical mechanics ◒; solar physics ◒ ③ Liquid crystals ○; nanomaterials ○; statistical mechanics ◒

Dundee ① Microelectronics ● ② Electromagnetic waves ◒; surface microbeam analysis ◒; integrated devices ◒

Durham ① Classical mechanics ●; statistical physics ● ② Classical mechanics ●

Edinburgh ① High performance computing in physics ●; physics on parallel computers ●; digital image analysis ○; atmospheric dynamics ○; fluid dynamics ○ ② Fluid dynamics ○; fundamental symmetries ○; atmospheric physics ○; quantum field theory ◒; macromolecular physics ○; advanced statistical physics ◒; groups and symmetries ○ ③ Atmospheric physics ○; fluid dynamics; digital image analysis ○

Exeter ① Computational physics ○; NMR ○; device physics ○; classical field theory ○; uses of ionising radiation ○; fluid mechanics ○; many body theory ○; semiconductors ○; quantum electronics ○ ② Uses of ionising radiation ●; electronics ○; computing ○; NMR ○ ③ Fluids ●; uses of ionising radiation ○ ④ Quantum electronics ●; communications ●; optoelectronic systems ●; energy and the environment ○ ⑤ Quantum electronics ●; communications ●; optoelectronic systems ●; energy and the environment ○; general relativity ○; fluid mechanics ○; device physics ○

Heriot-Watt ① Group theory ◒ ② Semiconductor devices ●; semiconductor technology ◒

Hertfordshire ① Advanced mechanics

Hull ① ② ③ ④ Business and professional skills ○

Imperial College London ① Research training ② Music ●; musical performance ●

Kent ① Space science ○; history of physical science ○; numerical and computational physics ○; classical physics ◒ ② Space science ○; history of physical science ○; classical physics ◒ ③ Space science ●; history of physical science ○; numerical and computational physics ○

Leeds ① Polymer physics ○; statistical mechanics ●; molecular simulation ○; medical imaging ○ ② Molecular astrophysics ●; star formation ●

Leicester ① Nuclear electronics ●; microprocessors ●; space science/technology ○; fluid mechanics ○; nuclear medicine ○ ② Nuclear electronics ●; microprocessors ●; space science/technology ○; fluid mechanics ○; cosmology ○ ③ ④ Nuclear electronics ●; microprocessors ●; space science/technology ○; fluid mechanics ○; Earth observation science ○

Liverpool ① Nuclear energy and environmental radiation ○; statistical mechanics ○ ② Nuclear energy and environmental radiation ○

Loughborough ① Wide range of engineering options ② Surface physics ○; fluid mechanics ◒; statistical mechanics ○
Manchester ① ② ③ ④ Meteorology ○
Nottingham ① Statistical physics ●; quantum statistics ◑ ② 2D physics ○; medical imaging ○; cosmology ○; atmospheric physics ○ ③ ④ Cosmology ● ⑤ Quantum statistics ●
Nottingham Trent ① Condensed matter ◒
Oxford ① Thermodynamics and statistical mechanics ●; physics of atmospheres and oceans ◒; laser science and quantum information processing ◒;
Queen Mary ① Statistical mechanics ○; chaotic dynamics ○
Reading ① Chaos theory
Royal Holloway ① Major project in applied physics ② Major astrophysics project ③ Further mathematics ○; non-linear phenomena and chaos ○; semiconductors and superconductors ○ ④ Further mathematics courses ○; non-linear phenomena and chaos ○; semiconductors and superconductors ○
Salford ① Computational physics; space technologies ② Communications in space ●; robotics in space ●; space medicine ◒
Sheffield ① Problem-solving in physics and astronomy ● ② ③ ④ Problem-solving in physics ●
Swansea ① Quantum devices ◒; physics of the body ○ ② Gravitational physics ●; particle physics and cosmology ●
York ① ② Thermodynamics ●; group theory ◒

MPhys and MSci courses

The majority of institutions now offer an MPhys or MSci course, which usually takes an extra year. TABLE 3a above shows which topics are only available if you take the MPhys course (◒); TABLE 3b lists the other differences between the Bachelor's and Master's degrees (note that the only courses featuring in this table are those that are available as both Bachelor's and Master's degrees).

Table 3b Differences between MPhys/MSci and BSc courses

Institution	Course title	Differences
Aberystwyth	Physics	Project occupies 50% of year 4; MPhys students examined to greater depth
Bath	Physics	Introduction to research with the department's research groups; specialist options relating to research interests of department; enhanced placements abroad and at research institutions in UK; full-semester project or 6-month placement
Belfast	Physics	At stage 3 MSci students take additional core modules and compulsory continuously assessed overview module instead of 1 term of practical physics; in final year there is substantial project work in lasers, condensed matter or atomic and molecular physics
	Physics with astrophysics	At stage 3 MSci students take additional core modules and compulsory continuously assessed overview module instead of 1 term of practical physics; in final year there is substantial project work in astrophysics
	Theoretical physics	Much more project work, particularly in year 4; MSci modules more closely specified
Birmingham	*All courses*	More year 3 courses are compulsory
Bristol	Physics	Wider range of options; in-depth project; option to specialise in one area of physics
	Physics with astrophysics	More depth and more training in research methods
Central Lancashire	*All courses*	More advanced topics in year 4; 1-semester full-time research project
Dundee	Physics	Project covers 2 years
Durham	Natural sciences (physics)	Major research project in year 4
	Physics	More options; year 4 in-depth research project contributes half of year's assessment
	Physics and astronomy	More options; year 4 in-depth research project contributes half of year's assessment
Edinburgh	*All courses*	Additional course options; group project in penultimate year; research project in final year
Exeter	*Both courses*	Advanced courses; research project in years 3 and 4 within one of the school's research groups

(continued) **Table 3b** Differences between MPhys/MSci and BSc courses

Institution	Course title	Differences
Glasgow	Astronomy	More advanced material and applications
	Physics	More advanced topics in year 4; 1-semester research project
Heriot-Watt	Computational physics	More advanced subjects; less vocational than BSc; intended for students aiming for PhD or research in industry
	Engineering physics	More advanced subjects; less vocational than BSc; intended for students aiming for PhD or research in industry
	Physics	More advanced subjects; less vocational than BSc; intended for students aiming for PhD or research in industry
Hull	Applied physics	Full-year project in final year; computer-aided physics; computer project; greater option choice
	Physics	Full-year project in final year; computer-aided physics; computer project; greater option choice
	Physics with lasers and photonics	Full-year project in final year; computer-aided physics; computer project; greater option choice
	Physics with medical technology	Full-year project in final year; computer-aided physics; computer project; greater option choice
Imperial College London	Physics	Substantial year 4 project; compulsory research interface course teaching financial, managerial and communications skills; year 4 options at enhanced level, many shared with Master's courses
	Physics with theoretical physics	Substantial year 4 project in theoretical physics; compulsory research interface course; year 4 options at Master's level, many shared with MSc in quantum fields and fundamental forces
Keele	*Both courses*	Extended research project and professional skills
Kent	Physics	MPhys students spend year 3 project time in preparation for their extended project in year 4
	Physics with astrophysics	MPhys students spend year 3 project time in preparation for their extended project in year 4
	Physics with space science and systems	MPhys students spend year 3 project time in preparation for their extended project in year 4
King's College London	Physics	MSci areas covered to higher level; 2:1 required in first 3 years to proceed to MSci
Leeds	Physics	Year 3 project may be for 1 semester abroad; wider range of options; year 4 project carried out in research group over longer period
	Physics with astrophysics	Year 3 project may be for 1 semester abroad; wider range of options; year 4 project carried out in research group
	Theoretical physics	Theoretical physics project in year 4
Leicester	Physics	Research level option courses; specialised directed reading study; research project; ABB A-level grades required
	Physics with astrophysics	Research level option courses; specialised directed reading study; research project; BBC A-level grades required
	Physics with planetary science	Research level option courses; specialised directed reading study; research project; BBC A-level grades required
	Physics with space science and technology	Research level option courses; specialised directed reading study; research project; BBC A-level grades required
Liverpool	Astrophysics	Extensive project work based in Astrophysics Research Institute, year 4
Loughborough	*Both courses*	Half of final year spent on full-time experimental or theoretical research project at Loughborough or in industry/research institution; salary of £1,000 per month may be payable
Manchester	*All courses*	Greater depth and breadth; extended projects; skills project
Nottingham	Chemistry and molecular physics	Provides grounding for PhD
	Mathematical physics	Year 4: substantial elements involving student-centred activities
	Physics	Year 3: 25% course consists of modules available to MSci students only; year 4: all work project/coursework, continuous assessment

(continued) **Table 3b** Differences between MPhys/MSci and BSc courses

Institution	Course title	Differences
Nottingham (continued)	Physics with astronomy	Year 3: 25% course consists of modules available to MSci students only; year 4: all work project/coursework, continuous assessment
	Physics with medical physics	Year 3: 25% course consists of modules available to MSci students only; year 4: all work project/coursework, continuous assessment
	Physics with theoretical astrophysics	More project work
	Physics with theoretical physics	Year 4: student-centred activities
Nottingham Trent	Physics	Larger industrial project in place of academic project in year 3; research and development methodology; industrial case studies and financial modules; slightly higher level of mathematics required
Oxford	Physics	2 major options of 40 hours of lectures each; full-time project occupying most of 1 term
Queen Mary	*All courses*	Higher-level treatment of subject; final year covers topics leading to research frontiers; higher entry requirement
Reading	Physics	Increased emphasis on research projects (eg computational, experimental); final-year content very different
	Theoretical physics	More project work; very different material in final year
Royal Holloway	Physics	More specialised options; more critical and independent learning; BBC required for entry
	Theoretical physics	More specialised options; more critical and independent learning; BBC required for entry
St Andrews	Astrophysics	Higher-level treatment of some topics; many topics only available for MPhys students; longer project
	Physics	Higher-level treatment of some topics; many topics only available for MPhys students; longer project
Salford	Physics	Substantial individual research project
	Physics with acoustics	Substantial individual research project
	Physics with space technology	Substantial individual research project
Sheffield	Physics	Design study project; advanced lecture courses; compulsory programming course; extended research project
	Physics with medical physics	Hospital or industrial placement; advanced lecture courses; compulsory programming course; extended research project
	Theoretical physics	Directed theoretical reading; advanced computational physics; advanced lecture courses; extended research project
Southampton	*All courses*	More demanding final-year project (usually linked to superviser's research); specialist options informed by cutting-edge research
Strathclyde	Applied physics	Greater depth; entrance requirements BBB at A-level, ABBB SQF
	Physics	Greater depth; more project work; entrance requirements BBB at A-level, ABBB SQF
Surrey	*All courses*	Full year of research spanning semesters 6 and 7, which may be taken at leading research establishment in UK, Europe or USA
Sussex	Physics	Advanced laboratory work in year 3
	Theoretical physics	3 half-term projects in year 3
Swansea	Physics	Extra laboratory courses in year 3; 1-semester research project or industrial placement in year 4
	Theoretical physics	1-semester research project in year 4
UCL	All courses	Group project in year 3; year 4 has wide choice of advanced options, including some given by other colleges in the University of London
Warwick	Physics	Greater depth; includes research element
York	*All courses*	More substantial final-year project; additional optional courses in final year

Mathematics and supporting/subsidiary content In all courses you will need to study one or more other subjects in addition to physics. Some of these subjects, such as mathematics, are taught to support your work in physics; some, such as foreign languages, are designed to teach you additional skills; while others may be there just to broaden your horizons. The subjects offered, the time they occupy and whether they are compulsory or optional varies considerably from institution to institution. TABLE 3c gives an indication of the pattern for courses listed in this Guide. For full information consult prospectuses or institutions directly. Institutions that run modular degree schemes can be particularly flexible with the topics they offer: see TABLE 2a for courses that are part of modular degree schemes in which you are relatively free to choose topics from a large number of different subjects.

Because of the importance of mathematics in physics courses, TABLE 3c gives more detailed information about the amount and level of mathematics included. Note that this data depends on some rather subjective judgements, and not all universities have supplied information – for these columns, a blank entry means no information was supplied, not that no mathematics is taught. Where information has been supplied, TABLE 3c shows the proportion of time spent taking specialised mathematics courses in the first, intermediate and final years of the course. In addition to this time, you may also study some mathematics within physics topics themselves. TABLE 3c also gives an indication of where the course lies in the range from the least mathematical to the most mathematical. A single course can spread across several columns because the level of mathematics depends on which options you choose. For example, options in theoretical physics will usually include more advanced mathematics.

Table 3c Mathematics and supporting/subsidiary content

Institution	Course title	Proportion of time spent on mathematics			Level of mathematics					Computing	Electronics	Engineering/technology	Other physical science	Biological science subjects	Modern language	Social science	Humanities	Music/arts
		First year %	Intermediate year %	Final year %	Least mathematical				Most mathematical									
Aberdeen	Physics									◑	◑	○	○	○	○		○	○
Aberystwyth	Physics	33	16	5	■		■			●								
	Physics with planetary and space physics									●								
Bath	Physics	20	20	10		■	■		■	●	●		○		○	○		
Belfast	Physics									○			○	○	○			
	Physics with astrophysics									○			○	○				
Birmingham	Physics									●	◑	○	○	○	○			
	Physics and astrophysics									●	◑				○			
	Physics and space research									●	◑				○			
	Theoretical physics									◑	○				○			
	Theoretical physics and applied mathematics									●								

Mathematics and supporting/subsidiary content

Institution	Course title	Proportion of time spent on mathematics: First year %	Intermediate year %	Final year %	Level of mathematics (Least mathematical – Most mathematical)	Computing	Electronics	Engineering/technology	Other physical science	Biological science subjects	Modern language	Social science	Humanities	Music/arts
Bristol	Physics					◑	●	○	○	○	○		○	○
	Physics with astrophysics					◑	●							
Cambridge	*Both courses*	25							○	○				
Cardiff	Astrophysics					●	●				○			
	Physics					●	●				○			
	Physics with astronomy					●	●				○			
	Physics with medical physics					●	●				○			
	Theoretical and computational physics	20			■	●	●				○			
Central Lancashire	Applied physics	17	17	0	■	●	●	○	○	○	○	○	○	○
	Astrophysics	17	17	0	■	●	○	○	○	○	○	○	○	○
	Mathematics and astronomy	0	0	0	■									
	Physics	17	17	0	■	●	●	○	○	○	○	○	○	○
Dundee	Applied physics	33	16	0	■	●	●							
	Physics	33	16	0	■									
Durham	Natural sciences (physics)	33	0	0	■	○			○	○	○	○	○	○
	Physics	33	16		■■■	●	●	○	○	○	○	○	○	○
	Physics and astronomy	33	16		■■■■	●	●	○	○	○	○	○	○	○
	Theoretical physics	33	16		■■■	●	●	○	○	○	○	○	○	○
Edinburgh	Astrophysics					○	○	○	○	○	○	○	○	○
	Computational physics					●	○	○	○	○	○	○	○	○
	Mathematical physics					○	○	○	○	○	○	○	○	○
	Physics					○	○	○	○	○	○	○	○	○
Exeter	Physics	16	16	0	■■■	◑	○	○	○	○	○	○	○	○
	Physics with astrophysics	6	16	0	■■■	◑	○	○	○	○	○	○	○	○
	Physics with medical applications	16	16	0	■	○	●			●	○			
	Physics with medical physics	16	16	0	■■	○	●			●				
	Physics with quantum and laser technology	16	16	0	■	○	●	○	○	○	○	○	○	○
	Quantum science and lasers	16	16	0	■■	○	●	○	○	○	○	○	○	○
Glasgow	Astronomy					●	○		○					
	Physics					●	●	○	○	○				
Heriot-Watt	Computational physics	25	25	0		●	●	○	○	○	○			
	Engineering physics	25	25	0	■	●	●	●	●	○	○			
	Physics	25	25	0		●	●	○	●	○	○			
Hertfordshire	Astronomy					●		○	○	○	○			○
	Astrophysics					●		○	○		○			○
Imperial College London	Physics	35	10	10	■	●	●	○	●	○	○		○	
	Physics with studies in musical performance	25	8	5	■	●	●	○	●	○	○		○	●
	Physics with theoretical physics	40	30	30	■	●	●	○	●	○	○		○	
Keele	Astrophysics	13	0	0	■	●	●							
	Physics	13	0	0	■	●	●		○					
Kent	Physics					●								
	Physics with astrophysics					●	○				○			

(continued) Table 3c

Mathematics and supporting/subsidiary content

Institution	Course title	Proportion of time spent on mathematics: First year %	Intermediate year %	Final year %	Level of mathematics (Least mathematical – Most mathematical)	Computing	Electronics	Engineering/technology	Other physical science	Biological science subjects	Modern language	Social science	Humanities	Music/arts
Kent (continued)	Physics with space science and systems					●					○			
King's College London	Mathematics and physics with astrophysics	50	50	50	■				●					
	Physics	15	25	25	■■									
	Physics with astrophysics	15	25	25	■■									
	Physics with medical applications	15	25	25	■									
Leeds	Physics					●	●	○	○	○	○	○	○	○
	Physics with astrophysics					●	●	○	○	○	○	○	○	○
	Theoretical physics					●	○	○	○	○	○	○	○	○
Leicester	Physics				■■■■	●	●				○			
	Physics with astrophysics				■■■■	●	●				○			
	Physics with planetary science					●	●				○			
	Physics with space science and technology					●	●				○			
Liverpool	Astrophysics					●			○					
	Mathematical physics					○	○							
	Physics					○	○		○		○			
	Physics for new technology						●							
Liverpool John Moores	Astrophysics	20	10		■	●								
	Physics with astronomy	20	15		■	●								
Loughborough	Engineering physics	17	17	0	■	◑	◑	◑	○		○			
	Physics	17	17	0	■	◑	◑	○	○		○			
Manchester	Physics	20	10		■■■■■■	◑	◑	○	○	○	○	○	○	○
	Physics with astrophysics	20	10		■■■■	◑	○	○	○	○	○	○	○	○
	Physics with technological physics	20	10		■■■	○	◑	◑	○	○	○	○	○	○
	Physics with theoretical physics	30	20		■■	◑	○	○	○	○	○	○	○	○
Nottingham	Chemistry and molecular physics					◑			○		○			
	Mathematical physics	50	50	50		●								
	Physics					●	●	○	○	○	○	○	○	○
	Physics with astronomy					●	●							
	Physics with medical physics					●	●			●				
Nottingham Trent	Physics					●	●		○	○	○			
Oxford	Physics	45	25	0	■	●	●				○			
Paisley	Physics					●	○		○		○			
Queen Mary	Astronomy					◑	◑	○	○		○			
	Astrophysics					◑	◑	○	○		○			
	Physics					●	○	○	○		○		○	○
	Theoretical physics					◑	○	○	○		○		○	○
Reading	Physics					●	●				○			
	Theoretical physics					●								

(continued) Table 3c Mathematics and supporting/subsidiary content

Institution	Course title	Proportion of time spent on mathematics: First year %	Intermediate year %	Final year %	Level of mathematics: Least mathematical					Most mathematical	Computing	Electronics	Engineering/technology	Other physical science	Biological science subjects	Modern language	Social science	Humanities	Music/arts
Royal Holloway	Applied physics										○	○		○		○			
	Astrophysics										○	○		○		○			
	Physics										○	○		○		○			
	Theoretical physics										○	○		○		○			
St Andrews	Astrophysics	33	20	0			■				○	○		○		○	○	○	
	Physics	33	20	0			■				●	●		○		○			
	Physics with photonics	33	20	0			■				●	●		○		○	○	○	
	Theoretical physics	33	20	0					■		○			○		○	○	○	
Salford	Physics										●	●							
	Physics with acoustics										●	●							
	Physics with space technology										●	●	●						
Sheffield	Astronomy	33	8	0			■				◑								
	Physics						■				◑	○	○	○	○	○	○	○	○
	Physics with medical physics	33	8	0			■				●	●	●						
	Theoretical physics								■		●	○	○	○	○	○	○	○	○
Southampton	Physics	25	0	0							○	○	○	○	○	○	○	○	○
	Physics with astronomy	25	0	0															
	Physics with photonics	25	0	0							○	○	○	○	○	○	○	○	○
	Physics with space science	25																	
Strathclyde	Applied physics	25	15	0		■					○	○	○	○	○	○	○	○	○
	Biophysics	25	15	0		■					○	●	○	○	●	○	○	○	○
	Laser physics and optoelectronics	25	15	0			■				○	○	○	○	○	○	○	○	○
	Photonics	30	15	0			■				●	○	○	○	○	○	○	○	○
	Physics	25	15	0			■				○	○	○	○	○	○	○	○	○
	Physics with visual simulation	25	15	10				■			●	○	○	○	○	○	○	○	○
Surrey	Physics										●	●				○			
	Physics with medical physics										●	●			●	○			
	Physics with nuclear astrophysics										●	●				○			
	Physics with satellite technology										●	●				○			
Sussex	Physics										◑	◑		◑		○	○	○	○
	Theoretical physics										◑	○		○		○	○	○	○
Swansea	Physics	25	0	0					■		●	●				○			
	Physics with nanotechnology	25	0	0				■			●	●							
	Physics with particle physics and cosmology	25	0	0					■		●	●							
	Theoretical physics	25	0	0						■									
Warwick	Physics	20	15	0							○	●	○		○	○	○		○
York	Physics										◑	●			○	○			
	Physics with astrophysics										●	●							
	Theoretical and computational physics										●				○	○			

Sandwich courses and other industrial experience Many courses provide a range of opportunities for spending a period of industrial training in the UK or abroad (more information about work and study abroad is given in the next section).

Several institutions offer a choice between a full-time course and a sandwich course: see TABLE 2a. Sandwich courses offer the advantage that after one or more industrial training periods you should be better able to relate theoretical concepts to industrial practice. You will also have a much better feel for the social and cultural aspects of working in industry, and, hopefully, you will have proved that you can work productively in an industrial environment. You may also have made some valuable contacts that will make your job search easier.

On the other hand, sandwich courses take longer than full-time courses. Taking a sandwich course can also affect your social life, as the extra time taken means that you will get 'out of phase' with other students on full-time courses, and during industrial periods you may lose contact with them.

Structure of sandwich courses There are several different ways of organising sandwich courses. There has traditionally been a division between 'thick' and 'thin' sandwich courses, though recently institutions have been moving away from offering thin sandwich courses and there are now comparatively few left. Thick sandwich courses have one or two industrial periods lasting about a year; thin sandwich courses have shorter periods, often of about six months, spent alternately in industry and university or college. Thin sandwich courses can offer the advantage of greater integration of academic work and industrial training, but there are more periods of disruption as you move from one environment to the other.

Thick sandwich courses can be divided into two main types: 2–1–1, in which there is a 12-month period of industrial training between the second and final years of the academic course, and 1–3–1, in which a three-year full-time degree course is 'sandwiched' between two one-year periods in industry.

Institutions usually arrange the periods of industrial experience, but you may be able to arrange your own, subject to approval.

Other forms of industrial experience Some institutions allow students to take an 'intercalated' year out in the middle of a full-time course. In some ways this is similar to a sandwich course, except that you are unlikely to be supervised by the university/college. TABLE 3d gives information about the opportunities for industrial experience for the courses in this Guide.

Time abroad Many institutions provide opportunities to spend time abroad in work experience or full-time study. A number of schemes exist for encouraging exchange between students in different countries. One of these is called Erasmus (European Community Action Scheme for the Mobility of University Students), which is part of the wider Socrates scheme.

A number of institutions run named variants of their main physics course that include time spent abroad as an integral part of the course. These have names like 'European physics' or 'Physics with a year in North America'. These courses normally have the same physics content as the main physics course at that institution, so they are not covered separately in the Guide: see prospectuses for details.

TABLE 3d gives information about the opportunities for spending time abroad while taking the courses in this Guide.

Table 3d Time abroad and sandwich courses

Institution	Course title	Location: ● Europe	○ North America	◒ industry	◑ academic institution	Maximum time abroad (months)	Socrates-Erasmus available	France	Germany	Italy	Spain	Netherlands	① ② ③ See footnotes	Sandwich courses: ● thick	○ thin	Arranged by: ● institution	○ student	Intercalated industrial year possible
Aberdeen	Natural philosophy	●	○		◑			●	●			●	①					
	Physics	●	○		◑		●	●	●		●	●	②					
Aberystwyth	*Both courses*	●	○	◒	◑	12	●	●	●	●	●	●	①					●
Bath	Physics	●	○	◒	◑	12	●	●	●					●	○	●	○	
Belfast	Physics	●			◑	12	●	●	●									
Birmingham	Physics	●			◑	9	●	●	●	●	●							
	Physics and astrophysics	●			◑	9	●	●	●	●	●							
	Theoretical physics	●			◑	9	●	●	●	●	●							
Bristol	Physics	●				12	●	●	●	●	●		①					
Central Lancashire	Applied physics	●	○	◒	◑	3	●	●	●	●	●							●
	Astrophysics	●	○		◑	3	●	●	●	●								●
	Physics	●	○	◒	◑	3	●	●	●	●	●							●
Dundee	*Both courses*		○			12												
Edinburgh	Astrophysics	●	○		◑	12		●										
	Computational physics	●	○		◑			●										
	Mathematical physics	●	○		◑	12	●	●		●								
	Physics	●	○		◑	12	●	●										
Exeter	Physics	●	○	◒	◑	12	●	●	●		●		①			●		●
	Physics with medical applications						●											●
	Physics with medical physics																	●
Glasgow	Astronomy	●	○		◑	12	●	●	●				①					
	Physics	●	○		◑	12	●	●	●				②					
Heriot-Watt	Computational physics					12	●	●	●									●
	Engineering physics	●				12	●	●										●
	Physics	●			◑	12	●	●	●									●
Hertfordshire	Astronomy	●	○	◒	◑	12	●	●	●	●	●	●	①	●		●	○	●
	Astrophysics	●	○	◒	◑	12		●	●	●	●	●	②	●		●	○	●
	Physics	●	○	◒	◑	12								●		●	○	
Hull	Applied physics	●	○			6	●		●				①					
	Physics	●	○			6	●		●				②					
	Physics with lasers and photonics	●	○			6	●		●				③					
	Physics with medical technology	●	○			6	●		●				④					
Imperial College London	Physics	●			◑	10	●	●	●	●	●		①					
Keele	*Both courses*	●	○		◑	6	●	●	●	●	●	●	①					

(continued) Table 3d Time abroad and sandwich courses

Institution	Course title	Location: ● Europe	○ North America	◗ industry	◑ academic institution	Maximum time abroad (months)	Socrates-Erasmus available	France	Germany	Italy	Spain	Netherlands	① ② ③ See footnotes	Sandwich courses: ● thick; ○ thin	Arranged by: ● institution	○ student	Intercalated industrial year possible
Kent	Physics	●	○			12	●	●	●	●	●						●
	Physics with astrophysics	●	○			12	●	●	●	●	●						●
	Physics with space science and systems	●	○			12	●	●	●	●	●						●
King's College London	Mathematics and physics with astrophysics	●	○			9	●	●	●	●	●	●					
	Physics	●	○			9	●	●	●	●	●	●					
	Physics with astrophysics	●	○			9		●	●	●	●	●					
	Physics with medical applications	●	○			9		●	●	●	●	●					
Lancaster	Physics		○		◑												
	Physics, astrophysics and cosmology		○		◑												
	Theoretical physics		○		◑												
Leeds	Physics	●	○		◑	9	●	●	●		●		①				
	Physics with astrophysics	●	○		◑	9	●	●	●		●		②				
Leicester	Physics	●			◑	12	●	●	●		●						
	Physics with astrophysics	●			◑	12	●	●	●		●						
	Physics with planetary science	●			◑	12		●	●								
	Physics with space science and technology	●			◑	12	●	●	●								
Liverpool	Physics					12	●	●	●		●		①				
Loughborough	*Both courses*	●	○	◗	◑	12	●	●	●	●	●	●	①	●	●	○	●
Manchester	Physics	●	○		◑	12	●	●	●	●	●	●	①				
	Physics with astrophysics		○		◑	12											
	Physics with technological physics		○		◑	12											
	Physics with theoretical physics		○		◑	12											
Nottingham	Chemistry and molecular physics	●				2	●	●									
	Physics	●	○		◑	10	●	●	●		●		①				
	Physics with astronomy		○		◑	10							②				
Nottingham Trent	Physics	●		◗		12	●	●	●					●	●		●
Paisley	Physics	●		◗		12	●	●	●					●	●	○	
	Technological physics					12								●	●	○	
Queen Mary	Astronomy	●				4	●	●	●	●	●		①				●
	Astrophysics	●				4	●	●	●	●	●		②				●
	Physics	●	○			12	●	●	●	●	●		③				●
	Theoretical physics	●				12	●	●	●	●	●		④				●
Royal Holloway	Physics	●			◑	12	●	●	●	●	●	●					
St Andrews	*All courses*		○		◑	9											
Salford	Physics	●	○	◗		12	●	●	●					●	●		●
	Physics with acoustics	●	○	◗	◑	12	●	●	●					●	●		●
	Physics with space technology	●	○	◗		12	●	●	●					●	●		●
Sheffield	Physics	●	○		◑	12	●	●	●		●						
	Theoretical physics		○		◑	12											
Southampton	Physics						●										
Strathclyde	Applied physics	●	○	◗	◑	12	●	●	●	●		●	①				
	Biophysics	●	○	◗	◑	12	●	●	●	●	●	●					
	Laser physics and optoelectronics	●	○	◗	◑	12		●	●	●		●					
	Photonics	●	○	◗	◑	12	●	●	●	●	●		②				
	Physics	●	○	◗	◑	12	●	●	●	●		●	③				

(continued) Table 3d

Time abroad and sandwich courses

Institution	Course title	Location: ● Europe	○ North America	◒ industry	◑ academic institution	Maximum time abroad (months)	Socrates-Erasmus available	France	Germany	Italy	Spain	Netherlands	① ② ③ See footnotes	Sandwich courses: ● thick; ○ thin	Arranged by: ● institution; ○ student	Intercalated industrial year possible
Strathclyde (continued)	Physics with visual simulation	●	○		◑	12	●		●	●	●					
Surrey	*All courses*	●	○	◒	◑	12	●	●	●			●	①	●	●	●
Sussex	Physics	●	○		◑	6	●						①			
	Physics with astrophysics	●				6	●						②			
	Theoretical physics	●			◑	6	●						③			
Swansea	Physics	●	○		◑	12	●	●	●	●			①			
Warwick	Physics	●	○	◒	◑	12	●	●	●	●	●	●	①			●
York	*All courses*	●	○		◑	9	●	●	●	●	●		①			

Aberdeen ①USA ②US
Aberystwyth ①All EU countries
Bristol ①Portugal; Sweden
Exeter ①USA; Australia; NZ
Glasgow ① ②USA; Canada
Hertfordshire ① ②Europe
Hull ① ② ③ ④Czech Republic (Prague)
Imperial College London ①Switzerland
Keele ①Scandinavia; USA; Canada
Leeds ① ②Denmark; Canada
Liverpool ①Belgium; Switzerland
Loughborough ①27 countries
Manchester ①Any European country
Nottingham ① ②Canada
Queen Mary ① ② ③ ④Belgium; Denmark; Portugal
Strathclyde ① ② ③Portugal
Surrey ①Sweden; Norway; Portugal
Sussex ① ② ③Sweden
Swansea ①Austria
Warwick ①All EU countries
York ①Portugal

Teaching methods Most courses are taught using a mixture of lectures, tutorials or supervisions (discussion groups with up to four or five students), and classes or seminars (larger discussion groups). The balance between these varies from one institution to another and often between the early and late stages of a course, with the emphasis switching from lectures to smaller group teaching.

Practical work Practical work is a major component of nearly all the courses and teaches important skills as well as providing backup to the theoretical content of lectures. The practical work can be organised in a variety of ways, but usually progresses from formal laboratory exercises to long-term project work. In many courses the practical work component is compulsory, and often your laboratory notes and reports are used as the basis for continuous assessment. You will be able to find out more about the organisation of practical work if you visit the institution and look around its teaching laboratories: this can be a good topic for discussion at an interview.

Computational techniques and IT skills are an important part of physics courses and are normally taught or supported through practical work in IT and computing laboratories.

Vacation courses and workshop practice Workshop techniques are taught at many institutions, often as part of laboratory training. Many institutions recommend or strongly encourage students to spend several weeks during one or more vacations working in industry or a government research laboratory, or possibly taking a course in workshop practice at the institution. Some institutions also arrange visits to industrial establishments, particularly for applied physics and sandwich courses.

Projects Projects play an important part in the assessment of many courses. They are usually carried out in the final year and give you the opportunity to pursue particular interests in greater depth, bringing together a range of knowledge and skills learnt during the course.

Projects can be either experimental or non-experimental. You usually have a fair amount of choice of topic, though you will receive advice from teaching staff, who will also have to approve your choice and then supervise the way you carry it out. Experimental projects may be drawn from any part of physics, subject to equipment availability, and may be carried out collectively by a group of students. Non-experimental projects may be computer based or a dissertation on some advanced topic. The length of the dissertation required varies, but might typically be around 15,000 words.

On many courses, you will need to give an oral presentation of what you have learned from the project in addition to a written report.

Students often find projects the most interesting and involving part of the course, though they can also be one of the most challenging. They can give you a taste of what being a real physicist is about.

Assessment methods Most institutions use a variety of assessment methods such as formal written examinations, continuous assessment of coursework and extended projects or dissertations. TABLE 4 gives information about the balance between these methods. It shows in which years there are written examinations and if they contribute to the final degree classification. However, the contribution from examinations in earlier years is often less than the contribution from the examinations in the final year. You should also note that although an examination may not contribute to the final result, passing it may be a condition for continuing with the course.

Many courses allow a wide range of options, which are often assessed in different ways, so it is difficult to give precise figures for the contributions of different assessment methods. For this reason, TABLE 4 shows the possible maximum and minimum contributions from coursework and projects or dissertations.

Frequency of assessment On many types of course, especially modular courses, assessment is carried out more frequently than in the traditional pattern of end-of-year examinations. Often, each module is assessed independently, soon after it has been completed. The precise details of when assessments are carried out vary from course to course: TABLE 4 shows if courses are assessed every term, semester (there are two semesters in a year) or year. Note that in modular courses, even if all modules are assessed, those occurring early in the course may not contribute as much to the final result as those later in the course.

The mix of assessment methods All assessment systems have advantages and disadvantages: for example, reducing the significance of final examinations may simply mean that short periods of high stress are replaced by a series of deadlines and continuous low-level stress throughout the course. Which of these you prefer will depend on your temperament. Because students vary in their response to different assessment methods, institutions usually employ a combination of methods, which also allows them to match the assessment method to the skill being tested. In some cases you may be able to change the make-up of your assessment regime, for example by choosing a dissertation or project instead of a formal examination.

Table 4 Assessment methods

Key to frequency of assessment: ● term; ◑ semester; ○ year

Institution	Course title	Frequency of assessment	Years of exams contributing to final degree (years of exams not contributing to final degree)	Coursework: minimum/maximum %	Project/dissertation: minimum/maximum %	Time spent on projects in final year (%) BSc	MPhys/MSci
Aberdeen	*Both courses*	◑	(1),(2),**3,4**	50/**60**	40/**50**	25	
Aberystwyth	*Both courses*	◑	(1),**2,3,4**	20/**30**	10/**20**	25	40
Bath	Physics	◑	(1),**2,3,4**	6/**24**	20/**36**	20	50

(continued) Table 4

Assessment methods

Key to frequency of assessment: ● term; ◑ semester; ○ year

Institution	Course title	Frequency of assessment	Years of exams contributing to final degree (years of exams not contributing to final degree)	Coursework: minimum/**maximum** %	Project/dissertation: minimum/**maximum** %	Time spent on projects in final year (%) BSc	MPhys/MSci
Belfast	Physics	◑	(1),**2,3,4**	20/**20**	20/**20**	30	45
	Physics with astrophysics	◑	(1),**2,3,4**	20/**20**	20/**20**	30	45
	Theoretical physics	◑	(1),**2,3,4**		5/**16**		
Birmingham	Physics	○	(1),**2,3,4**	20/**27**	17/**27**	30	33
	Physics and astrophysics	○	(1),**2,3,4**	20/**27**	17/**27**	30	33
	Physics and space research	○	(1),**2,3,4**	20/**27**	17/**27**	30	33
	Theoretical physics	○	(1),**2,3,4**	20/**27**	19/**19**		33
	Theoretical physics and applied mathematics	○	(1),**2,3,4**	10/**20**	19/**19**		33
Bristol	Physics	●	**1,2,3,4**	5/**5**	25/**25**	30	30
	Physics with astrophysics	●	(1),**2,3,4**	5/**5**	17/**25**	30	30
Cambridge	Natural sciences (astrophysics)	○	**3,4**	5/**10**	10/**25**		
	Natural sciences (physics, experimental and theoretical)	○	(1),(2),**3,4**	5/**10**	10/**25**		
Cardiff	Astrophysics	◑	(1),**2,3,4**		**15**	15	50
	Physics	◑	(1),**2,3,4**		**15**	15	50
	Physics with astronomy	◑	(1),**2,3,4**		**15**	15	20
	Physics with medical physics	◑	(1),**2,3**		**15**	15	
	Theoretical and computational physics	◑	(1),**2,3**		**15**	15	
Central Lancashire	Applied physics	◑	(1),**2,3,4**	40/**60**	17/**25**	17	50
	Astrophysics	◑	(1),**2,3,4**	40/**60**	17/**25**	17	50
	Mathematics and astronomy	◑		40/**60**	17/**25**	17	50
	Physics	◑	(1),**2,3,4**	40/**60**	17/**25**	17	50
Dundee	Applied physics	◑	(1),(2),**3,4**	15	20/**25**	25	
	Physics	◑	(1),(2),**3,4,5**	15	20/**25**	25	25
Durham	Natural sciences (physics)	○	(1),**2,3,4**	70/**75**	25/**30**	25	
	Physics	○	(1),**2,3,4**	70/**75**	25/**30**	25	50
	Physics and astronomy	○	(1),**2,3,4**	70/**75**	25/**30**	25	50
	Theoretical physics	○	(1),**2,3,4**	70/**75**	25/**30**		50
Edinburgh	Astrophysics	◑	(1),(2),**3,4,5**		30/**30**	22	38
	Computational physics	◑	(1),(2),**3,4,5**		30/**35**	25	35
	Mathematical physics	◑	(1),(2),**3,4,5**		5/**15**	5	25
	Physics	◑	(1),(2),**3,4,5**	**0**	30/**30**	25	35
Exeter	Physics	◑	(1),**2,3,4**	6/**7**	20/**23**	33	33
	Physics with astrophysics	◑	(1),**2,3,4**	6/**7**	20/**23**	33	33
	Physics with medical applications	◑	(1),**2,3**	7/**7**	20/**20**	33	
	Physics with medical physics	◑	(1),**2,3,4**	6/**6**	23/**23**		33
	Physics with quantum and laser technology	◑	(1),**2,3**	6/**7**	20/**23**	33	
	Quantum science and lasers	◑	(1),**2,3**	6/**7**	20/**23**		33
Glasgow	Astronomy	◑	(1),(2),**3,4**	17/**20**	13/**14**	16	16
	Physics	◑	(1),(2),**3,4**	13/**14**	8/**9**	16	16
Heriot-Watt	Computational physics	●	(1),(2),**3,4**	5/**8**	20/**20**	25	25
	Engineering physics	●	(1),(2),**3,4**	5/**8**	20/**20**		
	Physics	●	(1),(2),**3,4**	5/**8**	20	25	25
Hertfordshire	Astronomy	◑	(1),**2,3,4**	30/**50**	20/**20**		
	Astrophysics	◑	(1),**2,3,4**	30/**50**	20/**20**		

(continued) Table 4 Assessment methods

Key to frequency of assessment: ◐ term; ◑ semester; ○ year

Institution	Course title	Frequency of assessment	Years of exams contributing to final degree (years of exams not contributing to final degree)	Coursework: minimum/**maximum** %	Project/dissertation: minimum/**maximum** %	Time spent on projects in final year (%) BSc	MPhys/MSci
Hull	Applied physics	◑	(1),**2,3,4**	6/**10**	10/**20**	33	60
	Physics	◑	(1),**2,3,4**	6/**10**	10/**20**	33	60
	Physics with lasers and photonics	◑	(1),**2,3,4**	6/**10**	10/**20**	33	60
	Physics with medical technology	◑	(1),**2,3,4**	6/**10**	10/**20**	33	60
Imperial College London	Physics	○	**1,2,3,4**	5/**12**	7/**17**	15	25
	Physics with studies in musical performance	○	**1,2,3,4**	7/**10**	20/**25**	50	
	Physics with theoretical physics	○	**1,2,3,4**	7/**7**	10/**10**	25	30
Keele	*Both courses*	◑	(1),**2,3,4**	23/**23**	17/**17**	25	40
Kent	Physics	○	(1),**2,3,4**	15/**33**	15/**15**	12	38
	Physics with astrophysics	○	(1),**2,3,4**	15/**33**	15/**15**	12	38
	Physics with space science and systems	○	(1),**2,3,4**	15/**33**	15/**15**	12	38
King's College London	Mathematics and physics with astrophysics	◑	**1,2,3**	20/**20**	15/**15**	15	
	Physics	◑	**1,2,3,4**	20/**20**	15/**15**	15	25
	Physics with astrophysics	◑	**1,2,3**	20/**20**	15/**15**	15	
	Physics with medical applications	◑	**1,2,3**	20/**20**	15/**15**		
	Physics with molecular biophysics	◑	**1,2,3**	20/**20**	15/**15**	12	
Leeds	Physics	◑	(1),**2,3,4**	15/**20**	12/**23**	25	33
	Physics with astrophysics	◑	(1),**2,3,4**	15/**20**	12/**23**	25	33
	Theoretical physics	◑	(1),(2),**3,4**	**0**	8/**10**		25
Leicester	Physics	◑	(1),**2,3,4**	18/**33**	10/**20**	12	33
	Physics with astrophysics	◑	(1),**2,3,4**	18/**33**	10/**20**	12	33
	Physics with planetary science	◑	(1),**2,3,4**	18/**33**	10/**20**	12	33
	Physics with space science and technology	◑	(1),**2,3,4**	18/**33**	10/**20**	12	33
Liverpool	Astrophysics	◑	(1),**2,3,4**	12/**15**	12/**15**	10	20
	Mathematical physics	◑	(1),**2,3**	0/**10**	15/**25**		
	Physics	◑	(1),**2,3,4**	12/**15**	12/**15**	10	20
	Physics for new technology	◑	(1),**2,3**	15/**25**	10/**15**	10	
Liverpool John Moores	Astrophysics	◑	(1),**2,3,4**	10/**20**	25/**35**		30
	Physics with astronomy	◑	(1),**2,3**	10/**20**	15/**25**	20	
Loughborough	Engineering physics	◑	(1),**2,3,4,5**	10/**50**	25/**50**	25	50
	Physics	◑	(1),**2,3,4,5**	6/**25**	12/**24**	25	50
Manchester	Physics	◑	**1,2,3,4**	15/**30**	10/**30**	20	50
	Physics with astrophysics	◑	**1,2,3,4**	15/**30**	10/**30**	20	50
	Physics with technological physics	◑	**1,2,3,4**	15/**30**	10/**30**	15	40
	Physics with theoretical physics	◑	(1),**2,3,4**	15/**30**	10/**30**	20	50
Nottingham	Chemistry and molecular physics	◑	(1),**2,3,4**		16/**25**	10	30
	Mathematical physics	◑	(1),**2,3,4**	5/**10**	4/**10**	10	33
	Physics	◑	(1),**2,3**	15/**20**	15/**20**	17	35
	Physics with astronomy	◑	(1),**2,3**	15/**20**	15/**20**	17	35
	Physics with medical physics	◑	(1),**2,3**	15/**20**	15/**20**	17	35
Nottingham Trent	Physics	○	(1),**2,3,4**	41/**43**	18/**33**	33	25
Oxford	Physics	◐	(1),(2),**3,4**	**0**	8/**15**	16	33
Paisley	Physics	◑	(1),(2),(3),**4,5**	**0**	**25**	25	
Queen Mary	Astronomy	◑	(1),**2,3,4**	10/**20**	5/**20**	15	25
	Astrophysics	◑	(1),**2,3,4**	10/**20**	5/**20**	15	25
	Physics	○	(1),**2,3,4**	10/**20**	10/**20**	15	25

(continued) Table 4 Assessment methods

Key to frequency of assessment: ● term; ◑ semester; ○ year

Institution	Course title	Frequency of assessment	Years of exams contributing to final degree (years of exams not contributing to final degree)	Coursework: minimum/**maximum** %	Project/dissertation: minimum/**maximum** %	Time spent on projects in final year (%) BSc	MPhys/MSci
Queen Mary (continued)	Theoretical physics	○	(1),**2,3,4**	10/**20**	5/**20**	15	25
Reading	Physics	●	(1),**2,3,4**	35/**45**	8/**15**	20	40
	Theoretical physics	●	(1),**2,3,4**	35/**45**	8/**15**	20	40
Royal Holloway	Applied physics	○	**1,2,3,4**	10/**10**	15/**25**		25
	Astrophysics	○	**1,2,3,4**	10/**10**	15/**25**		25
	Physics	○	**1,2,3,4**	10/**10**	15/**25**	20	25
	Theoretical physics	○	**1,2,3,4**	10/**10**	15/**25**	20	25
St Andrews	Astrophysics	◑	(1),(2),**3,4,5**	10/**10**	8/**17**	25	50
	Physics	◑	(1),(2),**3,4,5**	10/**10**	8/**17**	25	50
	Physics with photonics	◑	(1),(2),**3,4,5**	10/**10**	17/**17**		50
	Theoretical physics	◑	(1),(2),**3,4,5**	10/**10**	12/**12**		37
Salford	Physics	◑	(1),**2,3,4,5**	7/**8**	25/**35**	33	46
	Physics with acoustics	◑	(1),**2,3,4,5**	7/**8**	25/**35**	33	46
	Physics with space technology	◑	(1),**2,3,4,5**	7/**8**	25/**35**	33	46
Sheffield	Astronomy	◑	(1),**2,3**	10/**20**	11/**11**	17	35
	Physics	◑	(1),**2,3,4**	10/**20**	11/**20**	17	33
	Physics with medical physics	◑	(1),**2,3,4**	10/**20**	11/**20**	17	33
	Theoretical physics	◑	(1),**2,3,4**	10/**20**	11/**20**	17	33
Southampton	Physics	◑	(1),**2,3,4**	40/**45**	15/**20**	25	25
	Physics with astronomy	◑	(1),**2,3,4**	40/**45**	15/**20**	25	25
	Physics with photonics	◑	(1),**2,3,4**	40/**45**	15/**20**	25	25
	Physics with space science	◑	(1),**2,3,4**	40/**45**	15/**25**	25	25
Strathclyde	Applied physics	◑	(1),(2),**3,4,5**	10/**20**	15/**20**	35	45
	Biophysics	◑	(1),(2),**3,4,5**	10/**20**	15/**30**		45
	Laser physics and optoelectronics	◑	(1),(2),**3,4**	10/**20**	15/**25**	35	
	Photonics	◑	(1),(2),(3),**4,5**	25/**30**	20/**25**		30
	Physics	◑	(1),(2),**3,4,5**	10/**20**	15/**20**	35	45
	Physics with visual simulation	◑	**4,5**	30/**40**	25/**35**		30
Surrey	Physics	◑	(1),**2,3,4**	10/**20**	15/**15**	25	25
	Physics with medical physics	◑	(1),**2,3,4**	10/**20**	15/**15**	25	50
	Physics with nuclear astrophysics	◑	(1),**2,3,4**	10/**20**	15/**15**	25	50
	Physics with satellite technology	◑	(1),**2,3,4**	10/**20**	15/**15**	25	50
Sussex	Physics		**1,2,3,4**	10/**15**	10/**15**	20	20
	Theoretical physics	○	**1,2,3,4**	5/**5**	10/**15**	20	20
Swansea	Physics	◑	(1),**2,3,4**	20/**25**	15/**40**	25	50
	Physics with nanotechnology	◑	(1),**2,3**	20/**25**	10/**15**	25	
	Physics with particle physics and cosmology	◑	(1),**2,3**	20/**25**	15/**20**	25	
	Theoretical physics	◑	(1),**2,3,4**	20/**25**	15/**40**	25	50
UCL	Astronomy	○	**1,2,3,4**	15	15	25	25
	Astrophysics	○	**1,2,3,4**	15	15	25	25
	Medical physics	○	**1,2,3,4**	15	15		25
	Physics	○	**1,2,3,4**	15	15	25	25
	Physics with medical physics	○	**1,2,3**	15	15	25	
	Theoretical physics	○	**1,2,3,4**		15/**25**	25	25
Warwick	Physics	○	**1,2,3,4**	20/**35**	10/**20**	25	25

(continued) **Table 4**

Assessment methods

Institution	**Course title** Key to frequency of assessment: ◐ term; ◑ semester; ○ year	**Frequency of assessment**	**Years of exams contributing to final degree** (years of exams not contributing to final degree)	**Coursework:** minimum/**maximum** %	**Project/dissertation:** minimum/**maximum** %	**Time spent on projects in final year** (%) BSc	MPhys/MSci
York	Physics	○	**1,2,3,4**	22/**24**	18/**22**	20	35
	Physics with astrophysics	○	**1,2,3,4**	22/**24**	18/**22**	20	35
	Theoretical and computational physics	○	**1,2,3,4**	18/**20**	18/**22**	20	35

How to apply Applications for courses in this Guide must be made online at www.ucas.com, the website of the Universities and Colleges Admissions Service (UCAS).

The UCAS tariff The UCAS points scheme, or 'tariff', combines GCE Advanced levels, GCE Advanced Subsidiary levels, AVCE and Scottish qualifications in a single points system. For example, a GCE A-level with an A grade or a Scottish Advanced Higher with an A grade will score 120 points, while an AS-level with an A grade, an A-level with a D grade or a Scottish Higher with a B grade will score 60 points.

For some courses, entrance requirements are expressed simply in terms of points, but many also make offers in terms of particular grades in A-levels or Highers, and some require minimum grades in specific A-levels or Highers. TABLE 5 summarises these requirements. However, for many courses the requirements are more complex, so you should take the information in the table only as a starting point, and an indication of the relative demands of the courses. It is vital to check prospectuses and university and college websites, to find out the full details of requirements for any courses you are considering.

Other qualifications All institutions accept certain other qualifications for their general requirement. For example, the International Baccalaureate (IB) and the European Baccalaureate (EB) are generally accepted. The publications listed in Chapter 6 give information about some other qualifications and about the acceptability of the IB and the EB for course-specific requirements. Again, if in doubt, you should consult institutions direct.

Institutions are becoming increasingly flexible about entrance requirements, so you should not assume that you will not be accepted if you lack standard qualifications. For example, in some franchising schemes you can begin at a level below the starting point of a degree course by taking the early stages of a course at a college of further education. Provided you are successful, you will then move on to an institution of higher education for the later stages of the course. You may also be interested in the Access courses offered at some universities and colleges, some of which are taught in the evenings. Successful completion of an Access course, while not guaranteeing you a place on a degree course, will give you an advantage when you apply.

Open days and interviews Many institutions and departments run open days, when you can look round their facilities and meet staff and students. In some cases you may be invited to visit only after you have been offered a place. For some courses, interviews play an important part in the selection process. They are usually conducted in as relaxed and friendly a manner as possible; the intention is to find out more about your personal qualities than can be learned from your application form alone.

Student numbers TABLE 5 also gives an estimate of the number of students on each course (the figure is shown in brackets if it is the total for a larger group, such as a faculty, or if it includes other students, such as those on a combined degree scheme, not just those studying physics). This number will give you some idea of how many other students you are likely to be taught with, but you should remember that some lectures may be common to several courses, and students on the same course may choose different specialised options, particularly in later years.

You should bear in mind that there is no simple equation between the number of places and the probability of acceptance, since a course with a large number of places may have those places because it is popular and oversubscribed. You cannot even use the ratio of applicants to places as a reliable guide, as one course may have relatively few applicants because it is known to have high entrance requirements, while another may attract a large number of applicants who view it as an insurance policy in case they fail to gain entry to another course.

Entrance requirements The information in TABLE 5 is for general guidance only, since admissions tutors consider applicants individually, and may take many factors into account other than examination grades.

Table 5 Entrance requirements

Institution	Course title	Number of students (includes other courses)	Typical offers: UCAS points	A-level grades	SCQF Highers grades	● compulsory; ○ preferred: A-level Physics	A-level Mathematics
Aberdeen	Natural philosophy	2		CCC	BBBB	●	●
	Physics	15	240	CDD	BBBB	●	●
Aberystwyth	Physics	15	240	BC		●	●
	Physics with planetary and space physics	15	240			●	●
Bath	Physics	85		ABC		●	●
Belfast	Physics	35		CCC	BBBC	●	●
	Physics with astrophysics	12		CCC	BBBC	●	●
	Theoretical physics	5		CCC	BBBC	●	●
Birmingham	Physics	65		ABB	AABBB	●	●
	Physics and astrophysics	40		ABB	AABBB	●	●
	Physics and space research	10		ABB	AABBB	●	●
	Physics with nanotechnology			AAB	AAABB		
	Physics with particle physics and cosmology			ABB	AABBB		
	Theoretical physics	10		ABB	AABBB	●	●
	Theoretical physics and applied mathematics	15		AAB	AAABB	●	●
Bristol	Physics	120		ABB	AAAAB	●	●
	Physics with astrophysics	40		ABB	AAAAB	●	●
Cambridge	*Both courses*	(600)		AAA		●	●
Cardiff	Astrophysics	30	260–300	BBC		●	●
	Physics	50	260–300	BBC		●	●
	Physics with astronomy	12	260–300	BBC		●	●
	Physics with medical physics	5		BBC		●	●

(continued) **Table 5**

Entrance requirements

Institution	**Course title**	**Number of students** (includes other courses)	**Typical offers** UCAS points	A-level grades	SCQF Highers grades	● compulsory; ○ preferred **A-level Physics**	**A-level Mathematics**
Cardiff (continued)	Theoretical and computational physics	2		BBC		●	●
Central Lancashire	Applied physics	(20)	280–320	BBC		●	●
	Astrophysics	(20)	280–320	BBC		●	●
	Mathematics and astronomy	10	240			○	○
	Physics	(20)	280–320	BBC		●	●
	Physics with astrophysics		280–320				
Dundee	*Both courses*	(25)	240	CCD		○	●
Durham	Natural sciences (physics)	201		AAA		●	●
	Physics	65		AAA		●	●
	Physics and astronomy	30		AAA		●	●
	Theoretical physics	20		AAA		●	●
Edinburgh	Astrophysics	50		BBB	BBBB	●	●
	Computational physics	5		BBB	BBBB	●	●
	Mathematical physics	12		ABC	ABBC	●	●
	Physics	50		BBB	BBBB	●	●
Exeter	Physics	50	280–300			●	●
	Physics with astrophysics	15	280–300			●	●
	Physics with medical applications	6	280			●	●
	Physics with medical physics	5	300			●	●
	Physics with quantum and laser technology	8	280			●	●
	Quantum science and lasers	5	280			●	●
Glamorgan	Astronomy	30	200–240				
Glasgow	Astronomy	58		BBC	BBBB	○	●
	Physics	(150)		BBC	BBBB	●	●
	Physics with astrophysics			BBC	BBB		
Heriot-Watt	Computational physics	12	200	CC	BBBC	●	●
	Engineering physics	6	200	CC	BBBC	●	●
	Mathematical physics		200		AABB		
	Nanoscience		200		AABB		
	Photonics and lasers		200		AABB		
	Physics	30	200	CC	BBBC	●	●
Hertfordshire	Astronomy	30	200–240			●	●
	Astrophysics	35	160–240			●	●
	Physics		160–240				
Hull	Applied physics	35	220–280		BCCCC	●	○
	Physics	35	240–300		BBBCC	●	●
	Physics with astrophysics		240–300				
	Physics with lasers and photonics	35	240–300		BBBCC	●	●
	Physics with medical technology	35	240–300		BBBCC	●	●
	Physics with nanotechnology		240–300				
Imperial College London	Physics	150		AAB		●	●
	Physics with studies in musical performance	3		AAB		●	●
	Physics with theoretical physics	50		AAA		●	●
Keele	Astrophysics	30	240–260		BBCC	●	○
	Physics	30	240–260		CCC	●	○
Kent	Astronomy, space science and astrophysics			BBC	AABBB		

Table 5 (continued) Entrance requirements

Institution	Course title	Number of students (includes other courses)	Typical offers: UCAS points	A-level grades	SCQF Highers grades	A-level Physics	A-level Mathematics
Kent (continued)	Physics			BBC	AABBB	○	○
	Physics with astrophysics			BBC	AABBB	○	○
	Physics with space science and systems			BBC	AABBB	○	○
King's College London	Mathematics and physics with astrophysics	4		AAB	AAAAB	●	●
	Physics	55		BBB	ABBBB	●	●
	Physics with astrophysics	5		BBB	ABBBB	●	●
	Physics with medical applications	5		BBB	ABBBB	●	●
	Physics with molecular biophysics	3		BBB	ABBBB	●	●
Lancaster	Physics	(70)		BBB	BBBBB	●	●
	Physics studies			CCC	AABB	●	○
	Physics with medical physics			BBB	BBBBB		
	Physics with particle physics and cosmology			BBB	BBBBB		
	Physics, astrophysics and cosmology			BBB	BBBBB	●	●
	Theoretical physics	(70)		BBB	BBBBB	●	●
	Theoretical physics with mathematics			ABB	AABBB		●
Leeds	Nanotechnology	10		BBB		●	●
	Physics	53		BBB		●	●
	Physics with astrophysics	20		BBB		●	●
	Theoretical physics	10		ABB		●	●
Leicester	Physics	20	280–320			●	●
	Physics with astrophysics	30	280–320			●	●
	Physics with nanoscience and technology		280–320				
	Physics with planetary science	30	280–320			●	●
	Physics with space science and technology	30	280–320			●	●
Liverpool	Astrophysics	20	320	ABC	ABBBC	●	
	Mathematical physics	10	320	BBB		●	●
	Physics	80	300–320	BBC	ABBBC	●	○
	Physics for new technology	10				●	○
	Physics with astronomy		300	BBC	BBBCC		
	Physics with medical applications		300	BBC	BBBCC		
	Theoretical physics		320				
Liverpool John Moores	Astrophysics	15	280–320			●	●
	Physics with astronomy	20	260–300	BBC		●	●
Loughborough	Engineering physics	20	280–300	BB	BB	●	●
	Physics	44	280–300	BB	BB	●	●
Manchester	Physics	175		AAA		●	●
	Physics with astrophysics	60		AAA		●	●
	Physics with technological physics	15		AAA		●	●
	Physics with theoretical physics	20		AAA		●	●
Nottingham	Chemistry and molecular physics	20		BBB		●	●
	Mathematical physics	25		AAB		●	●
	Physics	100		AAB		●	●
	Physics with astronomy	20		AAB		●	●
	Physics with medical physics	10		AAB		●	●
	Physics with theoretical astrophysics	20		AAB		●	●
	Physics with theoretical physics	20		AAB		●	●

● compulsory; ○ preferred

(continued) Table 5

Entrance requirements

Institution	Course title	Number of students (includes other courses)	Typical offers: UCAS points	A-level grades	SCQF Highers grades	A-level Physics (● compulsory; ○ preferred)	A-level Mathematics (● compulsory; ○ preferred)
Nottingham Trent	Astronomy		220				
	Physics	30	220–240			○	○
	Physics with astrophysics		220–260			○	○
	Quantum and cosmological physics		260–280				
	Technological physics		220				
Oxford	Physics	180		AAA		●	●
Paisley	Physics	(60)		DD	BBC	○	○
	Physics with medical technology			DD	BBC		
	Technological physics			DD	BBC		
Queen Mary	*All courses*	5	280–340		BBBBC	●	●
Reading	Physics	50	280–300			●	●
	Physics and the universe		280–300			●	●
	Theoretical physics	10	280–300			●	●
Royal Holloway	Applied physics	(20)	280–320	BBB		●	●
	Astrophysics	(20)	300	BBB	BBBBB	●	●
	Physics	(20)	280–300			●	●
	Physics with particle physics		260–300				
	Physics with quantum informatics			BBC			
	Theoretical physics	(20)	300–320			●	●
St Andrews	*All courses*	(90)		AAB	AAAB	●	●
Salford	Physics	(80)	180–260		CCCCC	●	●
	Physics with acoustics	(80)	180–260		CCCCC	●	●
	Physics with aviation studies		240	CCC		●	●
	Physics with pilot studies		240	CCC		●	●
	Physics with space technology	(80)	180–260		CCCCC	●	●
	Pure and applied physics		180–260				
Sheffield	Astronomy	25		ABB	AAAB	●	●
	Physics	70		ABC	ABBB	●	●
	Physics and astrophysics			ABC	ABBB	●	●
	Physics with medical physics	10		ABC	ABBB	●	●
	Theoretical physics	5		ABB	AABB	●	●
Southampton	Physics	40	340	ABB		●	●
	Physics with astronomy	20	340	ABB		●	●
	Physics with photonics	10	340	ABB		●	●
	Physics with space science	10	340	ABB		●	●
Strathclyde	Applied physics	30		BBC	BBCC	●	●
	Biophysics	8		BBB	ABBB	○	●
	Laser physics and optoelectronics	15		BBC	BBCC	●	●
	Photonics	8		BBB	ABBB	●	●
	Physics	45		BBC	BBCC	●	●
	Physics with visual simulation	6		BBB	ABBB	●	●
Surrey	Physics	25	260–300	BBB	BBBCC	●	●
	Physics with medical physics	10	260–300	BBB	BBBCC	●	●
	Physics with nuclear astrophysics	25	260–300	BBB	BBBBB	●	●
	Physics with satellite technology	20	260–300	BBB	BBBBB	●	●

(continued) Table 5 Entrance requirements

Institution	Course title	Number of students (includes other courses)	Typical offers: UCAS points	A-level grades	SCQF Highers grades	A-level Physics (● compulsory; ○ preferred)	A-level Mathematics (● compulsory; ○ preferred)
Sussex	Astrophysics			ABB	AABBB		
	Physics	20		ABB	AABBB	○	○
	Physics with astrophysics			BBC	BBBBB		
	Theoretical physics	4		BBC	BBBBB	●	●
Swansea	Physics	35	260–320		AABCC	●	●
	Physics with nanotechnology	10	260–280		AABCC	●	●
	Physics with particle physics and cosmology	25	260–280		AABCC	●	●
	Theoretical physics	10	300–320		AABBC	●	●
UCL	*All courses*	(60)		AAB		●	●
Warwick	Physics	150		AAB		●	●
York	*All courses*	(64)		BBB	BBBBB	●	●

Chapter 6: Professional organisations

The Institute of Physics is the professional body for physicists. The Institute provides, amongst other benefits, a forum for discussion for its members and a means of communication through its journals. By joining the Institute of Physics you will demonstrate a wider interest in and commitment to the subject, and membership can lead to chartered status, which is recognised across Europe as indicating a high level of professional competence.

The Institute of Physics offers three types of chartered status, depending on the nature of your work: Chartered Physicist (CPhys) for those working in physics; Chartered Engineer (CEng) for those working in engineering; and Chartered Scientist (CSci) for those working in more general, multidisciplinary science sectors. There are approximately 11,000 Chartered Physicists registered in the UK alone, and many more around the world. Chartered status requires an academic qualification (such as a BSc or MPhys) plus a period of work experience, which must be structured around some defined core skill requirements. Many companies accredit their graduate training programmes for CPhys or CEng to help new graduates achieve chartered status faster, and with a greater likelihood of success. The assessment for chartered status is rigorous, but it is well worth obtaining as it can often pay dividends later in your career. For further information about chartered status, go to www.iop.org and follow the link to Membership.

Chapter 7: Further information

Publications Unless indicated otherwise, all items in the following list are available from Trotman Publishing; phone 0870 900 2665, or buy online at www.trotman.co.uk (follow the link to 'Careers Portal', then to 'Your Bookshop').

The Big Guide 2008 UCAS, 2007, £29.50
How to Complete Your UCAS Application 2008 Entry Trotman, 2007, £11.99
Scottish Guide 2008 Entry UCAS, 2007, £14.95
The Ultimate University Ranking Guide C Harris. Trotman, 2004, £14.99
Student Book 2008 K Boehm & J Lees-Spalding. Trotman, 2007, £16.99
Experience Erasmus: The UK Guide Careerscope Publications, 2006, £16.95
Taking a Year Off Margaret Flynn. Trotman, 2002, £11.99
Students' Money Matters 2007 G Thomas. Trotman, 2007, £14.99
Mature Students' Directory Trotman, 2004, £19.99
Disabled Students' Guide to University E Caprez. Trotman, 2004, £21.99
Making the Most of University K van Haeften. Trotman, 2003, £9.99

Careers information

Careers with a Science Degree E D'Ath, T Doe & D Steel. Lifetime Careers, 2005, £10.99
Health Service Careers website www.nhscareers.nhs.uk; from 'Careers A–Z', follow the links to 'clinical engineer' and 'medical physicist'

Professional bodies

Institute of Physics Education Department, 76 Portland Place, London W1B 1NT; www.iop.org
Institute of Physics and Engineering in Medicine Fairmount House, 230 Tadcaster Road, York YO24 1ES; www.ipem.org.uk
Royal Astronomical Society Burlington House, Piccadilly, London W1J 0BQ; www.ras.org.uk

Physics websites

Physics.org www.physics.org
Eric Weisstein's World of Physics http://scienceworld.wolfram.com/physics
HyperPhysics http://hyperphysics.phy-astr.gsu.edu/hbase/hph.html
Physlink www.physlink.com
Physics Central www.physicscentral.com

General websites

Universities and Colleges Admissions Service www.ucas.com
Higher Education Funding Council for England www.aimhigher.ac.uk
Higher Education and Research Opportunities in the United Kingdom www.hero.ac.uk
Teaching Quality Information www.tqi.ac.uk; the site gives a range of information on courses, such as drop-out rates, student satifaction ratings and graduate destinations
Student Loans Company www.slc.co.uk

Student Support in Scotland www.student-support-saas.gov.uk
Student Support for Northern Ireland www.education-support.org.uk/students
Erasmus www.erasmus.ac.uk
European Commission Erasmus pages http://ec.europa.eu/; use the A–Z to select Education, then (from the menu) Programmes and Actions, then Erasmus
The European Choice: A Guide to Opportunities for Higher Education in Europe www.eurochoice.org.uk
Prospects Occupational Information (Higher Education Career Services Unit) www.prospects.ac.uk; from the 'Jobs and Work' menu, select 'Explore types of jobs'.
National Bureau for Students with Disabilities www.skill.org.uk

Background in physics This section lists a few of the very large number of books giving accounts of modern physics that are suitable for A-level students. They are not technical books and generally do not assume much specific knowledge, but they are trying to explain complex ideas so are not all easy to understand. However, they do give some of the flavour of what the subject is about and what it is like to be a physicist. Richard Feynman, Steven Weinberg and Stephen Hawking are amongst the most important contributors to theoretical physics in recent decades. Peter Medawar was a biologist, but his thoughts about science were not restricted to biology. Feynman, Weinberg and Medawar have all won the Nobel prize.

The periodicals *New Scientist* and *Scientific American* often contain articles of interest in the physics area. *New Scientist* is a weekly and is more concerned with 'news', while articles in *Scientific American* tend to be longer and are more often a review of a particular field. In addition to these, *Horizon* on television and *The Material World* on Radio 4 (see www.open2.net/materialworld) often cover issues of current interest in physics.

Black Holes, Wormholes and Time Machines J S Al-Khalili. Taylor & Francis, 1999, £14.99
A Short History of Nearly Everything B Bryson. Black Swan, 2004, £9.99
The Universe in a Nutshell S W Hawking. Bantam Press, 2001, £20.00
Advice to a Young Scientist P Medawar. Perseus, 1981, £21.00
The Limits of Science P Medawar. OUP, 1986
Cosmos C Sagan. Abacus, 1995, £9.99
Gravity From the Ground Up: an Introductory Guide to Gravity and General Relativity B Schutz. CUP, 2003, £40.00
The Flying Circus of Physics J Walker. Wiley, 2006, £14.95

The courses This Guide gives you information to help you narrow down your choice of courses. Your next step is to find out more about the courses that particularly interest you. Prospectuses cover many of the aspects you are most likely to want to know about, but some departments produce their own publications giving more specific details of their courses. University and college websites are listed in TABLE 2a, and can be reached via the UCAS site, www.ucas.com.

You can also write to the contacts listed below.

Aberdeen Student Recruitment and Admissions Service (sras@abdn.ac.uk), University of Aberdeen, Regent Walk, Aberdeen AB24 3FX

Aberystwyth Dr K Birkinshaw, Department of Physics, University of Wales, Aberystwyth SY23 3BZ

Bath Dr G Mathlin (g.mathlin@bath.ac.uk), Department of Physics, University of Bath, Bath BA2 7AY

Belfast Physics Physics with astrophysics Dr A Whitaker; Theoretical physics Dr D Findlay; both at Department of Pure and Applied Physics, The Queen's University of Belfast, Belfast BT7 1NN

Birmingham Dr Chris Mayhew (physics-adm@bham.ac.uk), School of Physics and Astronomy, University of Birmingham, Birmingham B15 2TT

Bristol Admissions Tutor (physics-admissions@bristol.ac.uk), H H Wills Physics Laboratory, University of Bristol, Tyndall Avenue, Bristol BS8 1TL

Cambridge Cambridge Admissions Office (admissions@cam.ac.uk), University of Cambridge, Fitzwilliam House, 32 Trumpington Street, Cambridge CB2 1QY

Cardiff Dr H Summers (summershd@cardiff.ac.uk), Department of Physics and Astronomy, Cardiff University, PO Box 913, Cardiff CF24 3YB

Central Lancashire Department Administrator (info4pasm@uclan.ac.uk), Department of Physics, Astronomy and Mathematics, University of Central Lancashire, Preston PR1 2HE

Dundee Dr D I Jones (d.i.jones@dundee.ac.uk), Electronic Engineering and Physics Division, University of Dundee, Dundee DD1 4HN

Durham Natural sciences (physics) Natural Sciences Admissions Tutor, Office of the Dean of Science; All other courses Admissions Tutor (physics.office@durham.ac.uk), Department of Physics; both at University of Durham, South Road, Durham DH1 3LE

Edinburgh Undergraduate Admissions Office (sciengug@ed.ac.uk), College of Science and Engineering, University of Edinburgh, The King's Buildings, West Mains Road, Edinburgh EH9 3JY

Exeter Tutor for Undergraduate Admissions (physug@exeter.ac.uk), School of Physics, University of Exeter, Stocker Road, Exeter EX4 4QL

Glamorgan Student Enquiry Centre (enquiries@glam.ac.uk), University of Glamorgan, Pontypridd, Mid Glamorgan CF37 1DL

Glasgow Astronomy Dr R Green (r.green@physics.gla.ac.uk); Physics Physics with astrophysics Dr A Watt (a.watt@physics.gla.ac.uk); both at Department of Physics and Astronomy, Glasgow University, Glasgow G12 8QQ

Heriot-Watt Computational physics Dr E Abraham (e.abraham@hw.ac.uk); Physics Dr M R Taghizadeh (m.taghizadeh@hw.ac.uk); All other courses Professor J I B Wilson (j.i.b.wilson@hw.ac.uk); all at Department of Physics, Heriot-Watt University, Edinburgh EH14 4AS

Hertfordshire J G Ling (admissions@herts.ac.uk), Director of Studies in Physical Sciences, University of Hertfordshire, College Lane, Hatfield AL10 9AB

Hull Dr J H C Hogg, Department of Physics, University of Hull, Hull HU6 7RX

Imperial College London Admissions Tutor (ph.admissions@imperial.ac.uk), Department of Physics, Imperial College London, South Kensington Campus, London SW7 2AZ

Keele Dr Roger Ward (r.p.ward@phys.keele.ac.uk), Admissions Tutor, School of Chemistry and Physics, Keele University, Staffordshire ST5 5BG

Kent Registry (recruitment@kent.ac.uk), University of Kent, Canterbury, Kent CT2 7NZ

King's College London Admissions Tutor (physics-secretary@kcl.ac.uk), Physics Department, King's College London, Strand, London WC2R 2LS

Lancaster Dr Tony Krier, Physics Department, Lancaster University, Lancaster LA1 4YB

Leeds Nanotechnology Andrew Nelson (nano@leeds.ac.uk); Physics with astrophysics Dr R A Duckett (r.a.duckett@leeds.ac.uk); Physics Theoretical physics Dr Stella Bradbury (physics.admissions@leeds.ac.uk); all at Department of Physics and Astronomy, University of Leeds, Leeds LS2 9JT

Leicester Dr G A Wynn (graham.wynn@astro.le.ac.uk), Department of Physics and Astronomy, University of Leicester, Leicester LE1 7RH

Liverpool Admissions Tutor (physics@liv.ac.uk), Department of Physics, University of Liverpool, PO Box 147, Liverpool L69 7ZE

Liverpool John Moores Student Recruitment Team (recruitment@ljmu.ac.uk), Liverpool John Moores University, Roscoe Court, 4 Rodney Street, Liverpool L1 2TZ

Loughborough Professor K R A Ziebeck (physics@lboro.ac.uk), Department of Physics, Loughborough University, Loughborough LE11 3TU

Manchester Dr F K Loebinger (fred.loebinger@manchester.ac.uk), School of Physics and Astronomy, Schuster Laboratory, University of Manchester, Manchester M13 9PL

Nottingham Chemistry and molecular physics Professor P J Sarre (peter.sarre@nottingham.ac.uk), School of Chemistry; Mathematical physics Mrs J Kenney (julie.kenney@nottingham.ac.uk); All other courses Professor M R Merrifield; both at School of Physics and Astronomy; all at University of Nottingham, University Park, Nottingham NG7 2RD

Nottingham Trent Admissions Office (sci.enquiries@ntu.ac.uk), School of Biomedical and Natural Sciences, Nottingham Trent University, Clifton, Nottingham NG11 8NS

Oxford Oxford Colleges Admissions Office, Oxford University, University Offices, Wellington Square, Oxford OX1 2JD

Paisley D G Robertson, Faculty of Science and Technology, University of Paisley, High Street, Paisley PA1 2BE

Queen Mary Dr W Gillin (w.gillin@qmul.ac.uk), Department of Physics, Queen Mary University of London, Mile End Road, London E1 4NS

Reading Dr Moira Hilton (physicsadmissions@rdg.ac.uk), Department of Physics, University of Reading, Whiteknights, Reading RG6 6AF

Royal Holloway E R Davies (e.r.davies@rhul.ac.uk), Admissions Tutor, Department of Physics, Royal Holloway, University of London, Egham, Surrey TW20 0EX

St Andrews Dr N C McGill (ncm@st-andrews.ac.uk), School of Physics and Astronomy, University of St Andrews, North Hough, St Andrews KY16 9SS

Salford Professor A D Boardman (a.d.boardman@salford.ac.uk), Department of Physics, Joule Laboratory, University of Salford, Salford M5 4WT

Sheffield Dr A M Fox (mark.fox@sheffield.ac.uk), Department of Physics and Astronomy, Hicks Building, University of Sheffield, Sheffield S10 2UN

Southampton Ms Kim Lange (entry@physics.soton.ac.uk), School of Physics and Astronomy, University of Southampton, Southampton SO17 1BJ

Strathclyde Dr Ronal Brown (ronal.brown@strath.ac.uk), Department of Physics, University of Strathclyde, Glasgow G4 0NG

Surrey Dr R Sear (p.bourke@surrey.ac.uk), Department of Physics, University of Surrey, Guildford GU2 7XH

Sussex Admissions Tutor (ug.admissions@physics.sussex.ac.uk), Physics and Astronomy, University of Sussex, Falmer, Brighton BN1 9QH

Swansea Dr C R Allton (physics@swansea.ac.uk), Admissions Tutor, Department of Physics, University of Wales Swansea, Singleton Park, Swansea SA2 8PP

UCL Dr Stan Zochowski (s.zochowski@ucl.ac.uk), Department of Physics and Astronomy, University College London, Gower Street, London WC1E 6BT

Warwick Dr M R Lees (m.r.lees@warwick.ac.uk), Department of Physics, University of Warwick, Coventry CV4 7AL

York Dr S D Tear, Department of Physics, University of York, York YO10 5DD

Part 2

Chapter 1: Introduction

Chemistry is one of the oldest sciences and now represents a vast quantity of accumulated knowledge and expertise, which no single person could possibly command. As a result, many diverse specialisms have grown up within the subject. Some of these, such as biochemistry and metallurgy, have become recognised as disciplines in their own right, while others are still thought of as part of the mainstream of chemistry.

A subject at the heart of science In branching out into different areas, chemists have made links with virtually all other areas of science. For example, chemistry is fundamental to understanding processes within living organisms, so chemists contribute to work across the full range of biological and medical studies. Much of the work of physical and theoretical chemists would not appear out of place in physics or mathematics departments. Geologists use inorganic chemistry to explain how rocks are formed and changed by conditions such as temperature, pressure and weathering. Environmental scientists use techniques from chemistry to detect and analyse environmental problems, and will need to rely on chemists to solve many of them.

However, there are also opportunities to focus on specific areas of chemistry, which are pursued for their own intrinsic interest and usefulness.

A subject with many uses Chemists make a vital contribution in many industries. The development of synthetic fibres, plastics, dyes, drugs and antibiotics, alloys, and, more recently, gas and dye lasers and micro-imaging techniques have all involved the practical application of chemical principles. All aspects of the food industry have been revolutionised by chemistry, from the fertilisers used by farmers and the additives used to preserve, flavour and stabilise processed food, to the analytical techniques used by public health authorities to monitor the end products. In fact, very few of the products we take for granted in our everyday life could be manufactured, packaged or distributed without the help of chemists at some stage.

Many services also rely on chemists. For example, you will find them working in hospital and forensic laboratories and for environmental agencies, as well as for private consultants working for industry.

A subject with a past The great advantage in studying a subject that has been around for as long as chemistry is that you have a huge armoury of theoretical and practical tools to draw on when you are tackling new problems. And as a student of chemistry – unlike some subjects – you can also be confident that you are building on firm foundations that have been tested by thousands of your predecessors.

A subject with a future However, you may be worried that a subject that has been around for hundreds of years will have become stale and worked out. Fortunately, nothing could be further from the truth: chemistry is still vigorous and exciting, and progress is being made now at a greater rate than at any other time. In modern chemistry, extensive use is made of a range of physical techniques, especially spectroscopic techniques for determining structure. Strong emphasis is placed on multinuclear magnetic resonance and mass spectrometry as well as infrared and ultraviolet spectroscopy. Computer modelling also features prominently, and this is directed at both the determination of molecular geometry and the elucidation of reaction profiles. There are still many unsolved and challenging problems requiring talented and imaginative minds, and investigating these problems can often open up whole new branches of the subject. The space available here allows mention of only a few examples of recent work, but these may give you a taste for the diversity, excitement and relevance of chemistry today.

New types of molecule New types of molecule are still being discovered, leading to work on their synthesis and investigations into their properties. For example, buckminsterfullerene is a molecule with 60 carbon atoms arranged in a cage-like spherical structure made up of 12 pentagons and 20 hexagons; it was named after Richard Buckminster Fuller who invented geodesic domes as architectural structures; it is also sometimes called soccerene because it looks like a football. Dendrimers are large complex branching polymers and are another new class of compounds exciting a great deal of interest (the Aldrich Chemical Co markets these under the trademark 'Starburst' compounds).

Alternative energy Chemists are making major contributions to the search for alternative sources of energy. They have an obvious research input into the exploitation of new fuels such as biomass and hydrogen, but are also key workers in solar energy, with contributions from the whole range of chemistry – physical, inorganic, organic and even theoretical. Approaches based on modelling photosynthesis form one line of attack, but there are many others.

Analysis Concerns for food safety and environmental protection are driving research into analytical techniques for detecting minute quantities of harmful substances and tracing them to their source.

Biological and medical research Some of the contributions to biological and medical research have been particularly important. Taxol is a powerful anti-cancer agent and, like many pharmaceuticals, was isolated from natural sources. Unfortunately, it only occurs naturally in the bark of one species of yew tree, making the extraction of useful quantities very difficult and expensive. However, chemists have been able to identify the features of the molecule that make it an effective drug, so that they can design new synthetic molecules with the same pharmacological action.

A variety of techniques drawn from chemistry, such as time-of-flight mass spectrometry, were used on the Human Genome Project, which in 2003 resulted in the complete mapping of some 30,000 genes, which make up the genetic blueprint of a human being. This has major implications for the understanding of the function of human genes and is likely to lead to significant medical benefits.

The Nobel Prize for Physiology and Medicine in 2003 was shared by Dr Paul Lauterbur, an American chemist, and Sir Peter Mansfield, a British physicist, for their discoveries in the field of magnetic resonance imaging (MRI), which is a medical imaging technique based on the principles of nuclear magnetic resonance (NMR). MRI is a powerful, non-invasive method used to diagnose various diseases and, more recently, to map brain function.

From A-level to degree chemistry Institutions are very aware that students come from having studied different A-level, AS-level or other courses, so an important goal of the first year of your degree course is to ensure that, whatever your background, by the end of the year you will have the knowledge and skills you need to meet the demands of the rest of the course.

There are a number of differences between the study of chemistry at school or college and at university. In most cases, attendance at lectures and classes, though highly recommended, is not compulsory, so you will be required to take much more responsibility for your own learning and study. In addition, you will be expected to undertake significantly more practical work than during your A-levels. You may find a first-year practical session in a large department with a hundred or more students a bit daunting, but you will receive considerable support to help you develop your practical techniques. You will also have the opportunity to use some highly sophisticated and advanced practical instruments, and later do a genuine research project.

Research Most of your teachers will be actively engaged in research, and discussions with them can be very stimulating and rewarding. Many institutions in the UK have justly gained an international reputation for the calibre of their research, and although there is no necessary correlation between the quality of the teaching in a department and the excellence of its research, most scientists would agree that there tend to be some parallels.

A career in chemistry Chemistry graduates have a wide choice of careers open to them. They can choose roles where a chemistry degree is essential, such as research and development jobs in the chemicals industry, or put their scientific training to use in a broader context, for instance as a patent agent or information specialist. Chemistry graduates may also opt for a career outside science, since their degree course encourages the development of transferable skills, such as analytical thinking and numeracy, that are sought after by many employers.

Further education and training Typically, almost as many chemistry graduates go on to study for some form of postgraduate qualification as enter employment directly. Postgraduate degrees can be divided into two categories: those that are largely taught and those that are mainly research-based. A taught course, such as a Master of Science (MSc) degree, will usually consist of a number of modules and will be assessed by a mixture of continuous assessment and exams. A research-based course, such as a Doctor of Philosophy (PhD) degree, consists of a substantial piece of individual, original research carried out under the supervision of a member of staff. The range of possible research topics is enormous.

A research qualification is virtually a prerequisite in some career areas, including many research posts in industry and universities. In addition, postgraduate qualifications, particularly those involving some research, prove a graduate's practical ability, which is of great importance for many employers. However, to be considered for a PhD, you will need at least a good upper second-class honours degree, or a Master's degree, and even then, funding can be difficult to obtain.

Who employs chemists? Many chemistry graduates continue to use chemistry in their subsequent work, and degree courses are designed to prepare students to meet the theoretical and practical demands that employers will place upon them. There are many industries in which you can use a chemistry degree directly: chemists are involved in the development of an enormous range and variety of products including pharmaceuticals, agrochemicals, dyes, paints, fibres, plastics, explosives, and toiletries such as toothpaste, soap and perfumes. Other chemists are employed in the metals, glass, electronics and food industries, and in utilities and energy companies. Opportunities also arise in research associations, health authorities and education.

The public sector, including the civil service and local government, is also an important recruiter of chemists. The following are some examples of the opportunities available. Public Health Laboratories analyse food that is processed and distributed to the public, as well as studying the spread of disease. Forensic science laboratories provide a service to police forces around the country. The Laboratory of the Government Chemist studies preservatives, flavourings and the chemistry of foods. The Health and Safety Executive deals with chemical safety, and the Environment Agency is concerned with the quality of water and environmental pollution. Information on careers within the civil service can be found at www.careers.civil-service.gov.uk.

What do professional chemists do? Research and development Chemists working for very different types of employer are often involved in similar types of work. Part of the appeal of research and development is that you have the chance to continue with practical laboratory work. Research chemists investigate ways of synthesising or isolating new chemicals for use as drugs, plastics, pesticides and a wide range of other products. Other research is focused on improving existing chemical processes. Development work involves finding the most efficient and economic means of producing chemicals on a large scale.

Research and development jobs are laboratory-based and demand creative flair. Employers usually require a good degree and it is worth bearing in mind that if you are interested in a career in research and development some companies prefer to recruit graduates with a four-year MChem/MSci qualification rather than a three-year BSc. Chemists employed in research and development usually work in a project team, normally made up largely of chemists, but which, depending on the project, may also include biochemists, microbiologists, pharmacologists or materials scientists.

Production The production phase usually follows development, and graduate chemists are employed to plan the reaction conditions, control the rate of production and supervise other production workers. Throughout this stage, other chemists use a variety of analytical procedures to assess the quality and purity of raw materials, reaction intermediates and finished products. Customers buying chemical products are likely to deal with chemists specialising in technical sales. In addition to promoting a product range, they offer advice on how to use the products, and discuss possible modifications, which are then relayed back to the research and development laboratories.

Support roles Chemists are also employed in jobs supporting the research, development and production teams. For example, manufacturers protect their new products by patenting them and may employ chemists as patent staff to write technico-legal specifications for their products. Some chemists are employed as information specialists to search the technical literature for information about chemicals and chemical processes for research and development staff, while others work as technical writers, producing company literature explaining their products to customers. All these occupations require considerable knowledge of and experience in chemistry. Employers will often require graduates to work on unfamiliar compounds or in areas not directly covered by their degree, so their ability to acquire new knowledge and develop appropriate skills is taken for granted. Chemists also enter the workplace at technician level in industry, education and the health service.

For more information on careers in chemistry, see the website of the Royal Society of Chemistry at www.rsc.org.

Teaching Opportunities for chemists in teaching are good as there is a national shortage of science teachers, and special incentives are offered to encourage scientists to enter teaching. For more information, see page 7 in the Physics section of this book.

Other opportunities Although many chemistry graduates use their degree directly in their employment, around 40% of all graduate vacancies are open to individuals regardless of their degree subject. This is because employers are prepared to train recruits and are more interested in their intellectual ability, experience and personal skills than specific subject knowledge. Chemistry students develop skills such as communication, organisation, teamwork and problem-solving through their studies. They also develop personal skills through work experience such as sandwich placements or vacation jobs, and participation in student sports and societies.

Chemistry graduates are well equipped to enter a variety of jobs outside the scientific arena. Possibilities include working in: journalism, publishing, accountancy, banking, sales, teaching, health and safety, forensics, law, the computing industry, food technology and retailing. Thus, a wide variety of opportunities are open to the chemistry graduate with good personal skills, both within and outside science.

Should you apply for a chemistry course?

Chemistry has widespread practical applications, making it a good preparation for a career in a wide variety of industries. However, it is also a challenging and exciting subject in its own right, and the study of chemistry is a satisfying and useful educational experience even if you do not intend to become a professional chemist.

Chapter 2: The courses

TABLE 2a lists the specialised and combined courses at universities and colleges in the UK that lead to the award of an honours degree in chemistry. You should refer to the notes preceding TABLE 2a in the first part of this Guide for guidance on using the table.

Table 2a First-degree courses in **Chemistry**

Institution Course title (① ② ③ see combined subject list – Table 2b)	**Degree**	**Duration** Number of years	**Foundation year** ● at this institution	○ at franchised institution	◑ second-year entry	**Modes of study** ● full-time	◗ part-time	○ time abroad	◐ sandwich	❂ **Modular scheme**	**Course type** ● specialised	◐ combined	**No of combined courses**
Aberdeen *www.abdn.ac.uk*													
Chemistry ①	BSc/MChem	4, 5			◑	●	◗	○		❂	●	◐	5
Chemistry with e-chemistry	MChem	5			◑	●					●		0
Medicinal chemistry	BSc/MChem	4, 5			◑	●	◗				●		0
Aston *www.aston.ac.uk*													
Applied chemistry ①	BSc/MChem	4, 5	●	○					◐	❂	●	◐	1
Chemistry ②	BSc/MChem	3, 4	●	○		●		○	◐		●	◐	13
Chemistry (biotechnology)	BSc	3, 4				●			◐		●		0
Chemistry (environmental management)	BSc	3, 4				●			◐		●		0
Chemistry (management studies)	BSc	3, 4				●			◐		●		0
Bangor *www.bangor.ac.uk*													
Chemistry	BSc/MChem	3, 4	●			●		○	◐	❂	●	◐	4
Bath *www.bath.ac.uk*													
Chemistry ①	BSc/MChem	3, 4	●			●		○	◐		●	◐	1
Chemistry for drug discovery	BSc/MChem	3, 4	●			●		○	◐		●		0
Belfast *www.qub.ac.uk*													
Chemistry ①	BSc/MSci	3, 4	●			●		○			●	◐	1
Chemistry with forensic analysis	BSc	3				●					●		0
Medicinal chemistry	BSc	3, 4				●			◐		●		0
Birmingham *www.bham.ac.uk*													
Chemistry ①	BSc/MSci/MNatSci	3, 4	●			●		○	◐		●	◐	5
Chemistry with analytical science	BSc/MSci/MNatSci	3, 4				●			◐		●		0
Chemistry with bio-organic chemistry	BSc/MSci/MNatSci	3				●			◐		●		0
Bradford *www.bradford.ac.uk*													
Chemistry with pharmaceutical and forensic science	BSc/MChem	3, 4	●			●			◐		●		0
Brighton *www.brighton.ac.uk*													
Pharmaceutical and chemical science	BSc	3				●	◗		◐	❂		◐	1
Bristol *www.bris.ac.uk*													
Chemical physics	BSc/MSci	3, 4				●		○	◐		●		0
Chemistry ①	BSc/MSci	3, 4	●			●		○	◐		●	◐	1
Bristol UWE *www.uwe.ac.uk*													
Biological and medicinal chemistry	BSc	3, 4	●			●	◗	○	◐		●		0
Forensic chemistry	BSc	3				●					●		0
Cambridge *www.cam.ac.uk*													
Natural sciences (chemistry)	BA/MSci	3, 4				●				❂	●		0
Cardiff *www.cardiff.ac.uk*													
Chemistry	BSc/MChem	3, 4	●			●		○	◐		●	◐	2
Dundee *www.dundee.ac.uk*													
Pharmaceutical chemistry	BSc	3, 4			◑	●		○			●		0

First-degree courses in **Chemistry**

Institution Course title (① ② ③ see combined subject list – Table 2b)	**Degree**	**Duration** Number of years	**Foundation year** ● at this institution	○ at franchised institution	◑ second-year entry	**Modes of study** ● full-time;	◒ part-time	○ time abroad	◐ sandwich	❂ **Modular scheme**	**Course type** ● specialised;	◐ combined	**No of combined courses**
Durham *www.durham.ac.uk*													
Chemistry	BSc/MChem	3, 4				●		○	◐		●		0
East Anglia *www.uea.ac.uk*													
Biological and medicinal chemistry	BSc/MChem	3, 4	●	○		●		○			●		0
Chemical physics	BSc/MChem	3, 4	●	○		●		○	◐		●		0
Chemistry	BSc/MChem	3, 4	●	○		●		○	◐	❂	●		0
Chemistry with analytical and forensic science	MChem	4	●	○		●					●		0
Chemistry with analytical science	BSc	3	●	○		●					●		0
Pharmaceutical chemistry	BSc	3	●	○		●					●		0
Edinburgh *www.ed.ac.uk*													
Chemical physics	BSc/MChem	5			◑	●					●		0
Chemistry ①	BSc/MChem	4, 5			◑	●		○	◐		●	◐	1
Chemistry with environmental chemistry ②	BSc/MChem	4, 5			◑	●		○	◐		●	◐	1
Medicinal and biological chemistry	BSc/MChem	4, 5			◑	●		○	◐		●		0
Exeter *www.exeter.ac.uk*													
Biological and medicinal chemistry	BSc	3, 4	●			●		○	◐		●		0
Glamorgan *www.glam.ac.uk*													
Chemistry	BSc	3, 4, 5	●			●	◒	○	◐	❂	●	◐	1
Glasgow *www.gla.ac.uk*													
Chemistry ①	BSc/MSci	4, 5			◑	●		○	◐		●	◐	3
Chemistry with medicinal chemistry	BSc/MSci	4, 5			◑	●		○	◐		●		0
Greenwich *www.gre.ac.uk*													
Analytical chemistry	BSc	3, 4, 5	●	○		●	◒	○	◐		●		0
Chemistry ①	BSc/MChem	3, 4, 5	●	○		●	◒	○	◐	❂	●	◐	7
Pharmaceutical chemistry	BSc	3, 4, 5	●	○		●	◒	○	◐		●		0
Heriot-Watt *www.hw.ac.uk*													
Chemistry ①	BSc/MChem	4, 5			◑	●		○	◐	❂	●	◐	9
Chemistry with pharmaceutical chemistry	BSc/MChem	4, 5			◑	●					●		0
Huddersfield *www.hud.ac.uk*													
Chemistry ①	BSc/MChem	3, 4	●			●	◒	○	◐		●	◐	4
Chemistry with analytical chemistry	BSc	3, 4	●			●	◒	○	◐		●		0
Chemistry with medicinal chemistry	BSc	3, 4	●			●		○	◐		●		0
Hull *www.hull.ac.uk*													
Chemistry ①	BSc/MChem	3, 4		○		●		○	◐		●	◐	1
Chemistry with analytical chemistry and toxicology	BSc/MChem	3, 4		○		●		○	◐		●		0
Chemistry with echem	BSc/MChem	3, 4				●		○	◐		●		0
Chemistry with forensic science and toxicology	BSc/MChem	3, 4		○		●		○	◐		●		0
Chemistry with molecular medicine	BSc/MChem	3, 4				●		○	◐		●		0
Chemistry with nanotechnology	BSc/MChem	3, 4				●		○	◐		●		0
Imperial College London *www.imperial.ac.uk*													
Chemistry ①	BSc/MSci	3, 4, 5				●		○	◐		●	◐	1
Chemistry with conservation science	MSci	4				●					●		0
Chemistry with fine chemicals processing	MSci	4, 5				●			◐		●		0
Chemistry with medicinal chemistry	MSci	4, 5				●		○	◐		●		0
Keele *www.keele.ac.uk*													
Chemistry ①	BSc/MSci/MChem	3, 4	●			●		○	◐	❂	●	◐	28
Medicinal chemistry ②	BSc	3	●			●		○	◐	❂	●	◐	28
Kent *www.ukc.ac.uk*													
Forensic chemistry	BSc	3, 4				●			◐		●		0

First-degree courses in **Chemistry**

Institution Course title (① ② ③ see combined subject list – Table 2b)	**Degree**	**Duration** Number of years	**Foundation year** ● at this institution	○ at franchised institution	◑ second-year entry	**Modes of study** ● full-time	◗ part-time	○ time abroad	◐ sandwich	✤ **Modular scheme**	**Course type** ● specialised	◐ combined	**No of combined courses**
Kingston *www.kingston.ac.uk*													
Chemistry ①	BSc/MChem	3, 4		○		●			◐	✤	●	◐	7
Chemistry (applied)	BSc/MChem	3, 4		○		●			◐	✤	●		0
Medicinal chemistry	BSc	3, 4		○		●			◐		●		0
Leeds *www.leeds.ac.uk*													
Applied chemistry	BSc	3		○		●				✤	●		0
Chemistry ①	BSc/MChem	3, 4		○		●		○	◐		●	◐	12
Chemistry with analytical chemistry	BSc/MChem	3, 4		○		●		○	◐		●		0
Colour and polymer chemistry	BSc	3, 4		○		●			◐	✤	●		0
Medicinal chemistry	BSc	3, 4				●		○	◐		●		0
Leicester *www.le.ac.uk*													
Chemical biology	BSc/MChem	3, 4		○		●		○	◐		●		0
Chemistry ①	BSc/MChem	3, 4		○		●		○	◐		●	◐	2
Pharmaceutical chemistry	BSc/MChem	3, 4		○		●		○	◐		●		0
Liverpool *www.liv.ac.uk*													
Chemistry ①	BSc/MChem	3, 4		○		●		○	◐	✤	●	◐	3
Medicinal chemistry	BSc	3		○		●					●		0
Medicinal chemistry with pharmacology	MChem	4		○		●					●		0
Liverpool John Moores *www.ljmu.ac.uk*													
Medicinal and analytical chemistry	BSc	3, 4	●			●		○	◐		●		0
Medicinal chemistry	BSc	3, 4	●			●		○	◐		●		0
Pharmaceutical science and biological chemistry	BSc	3, 4	●			●		○	◐		●		0
London Metropolitan *www.londonmet.ac.uk*													
Biological and medicinal chemistry	BSc	3, 4	●			●	◗	○	◐	✤	●		0
Chemistry ①	BSc	3, 4	●			●	◗	○	◐	✤	●	◐	4
Loughborough *www.lboro.ac.uk*													
Chemistry ①	BSc/MChem	3, 4, 5	●			●		○	◐		●	◐	2
Chemistry with analytical chemistry	BSc/MChem	3, 4, 5	●			●		○	◐		●		0
Chemistry with forensic analysis	BSc/MChem	3, 4, 5	●			●		○	◐		●		0
Medicinal and pharmaceutical chemistry	BSc/MChem	3, 4, 5	●			●		○	◐		●		0
Manchester *www.man.ac.uk*													
Chemistry ①	BSc/MChem	3, 4	●			●		○	◐	✤	●	◐	1
Chemistry with forensic and analytical chemistry	MChem	4				●					●		0
Chemistry with patent law	MChem	4				●				✤	●		0
Medicinal chemistry	BSc/MChem	3, 4				●					●		0
Manchester Metropolitan *www.mmu.ac.uk*													
Analytical chemistry	BSc/MChem	3, 4	●	○		●					●		0
Chemical science	BSc	4	●	○		●		○	◐		●		0
Chemistry ①	BSc/MChem	3, 4	●	○		●		○	◐		●	◐	23
Forensic chemistry	BSc	3	●	○		●					●		0
Medicinal and biological chemistry	BSc/MChem	3, 4	●	○		●		○			●		0
Pharmaceutical chemistry	BSc	4				●					●		0
Newcastle *www.ncl.ac.uk*													
Chemistry ①	BSc/MChem	3, 4	●	○		●		○	◐	✤	●	◐	3
Chemistry with medicinal chemistry	BSc/MChem	3, 4				●		○	◐	✤	●		0
Northumbria *www.northumbria.ac.uk*													
Applied chemistry	BSc/MChem	3, 4	●			●			◐	✤	●		0
Chemistry ①	MChem	4				●				✤	●	◐	1
Chemistry with forensic chemistry	BSc	3, 4				●			◐	✤	●		0

(continued) Table 2a

First-degree courses in **Chemistry**

Institution Course title	① ② ③ see combined subject list – Table 2b	**Degree**	**Duration** Number of years	**Foundation year** ● at this institution	○ at franchised institution	◑ second-year entry	**Modes of study** ● full-time	◡ part-time	○ time abroad	◐ sandwich	⊛ **Modular scheme**	**Course type** ● specialised	◐ combined	**No of combined courses**
Northumbria (continued)														
Pharmaceutical chemistry		BSc	3, 4				●			◐	⊛	●		0
Nottingham *www.nottingham.ac.uk*														
Chemistry	①	BSc/MSci	3, 4, 5				●		○	◐	⊛	●	◐	2
Medicinal and biological chemistry		BSc/MSci	3, 4				●					●		0
Nottingham Trent *www.ntu.ac.uk*														
Chemistry	①	BSc/MChem	3, 4				●		○	◐		●	◐	1
Chemistry with analytical science		BSc/MChem	3, 4				●			◐		●		0
Oxford *www.ox.ac.uk*														
Chemistry		MChem	4				●					●		0
Paisley *www.paisley.ac.uk*														
Chemistry	①	BSc	4, 5			◑	●	◡		◐		●	◐	2
Medicinal chemistry		BSc	4, 5				●			◐		●		0
Plymouth *www.plymouth.ac.uk*														
Chemistry (analytical)		BSc	3	●	○		●	◡	○		⊛	●		0
Chemistry (applied)		BSc	3	●	○		●	◡	○		⊛	●		0
Queen Mary *www.qmul.ac.uk*														
Pharmaceutical chemistry		MSci	4				●					●		0
Reading *www.rdg.ac.uk*														
Chemistry	①	BSc/MChem	3, 4	●			●		○	◐		●	◐	1
Chemistry with forensic analysis		BSc	3				●					●		0
Chemistry with medicinal chemistry		MChem	4	●			●					●		0
St Andrews *www.st-and.ac.uk*														
Chemical sciences		BSc	4			◑	●					●		0
Chemistry	①	BSc/MChem	4			◑	●	◡	○	◐	⊛	●	◐	9
Chemistry with catalysis		BSc	4			◑	●	◡	○	◐	⊛	●		0
Chemistry with materials chemistry		BSc	4			◑	●		○	◐	⊛	●		0
Chemistry with medicinal chemistry		BSc/MChem	4, 5			◑	●		○	◐		●		0
Sheffield *www.sheffield.ac.uk*														
Chemical physics		BSc/MChem	3, 4				●					●		0
Chemistry	①	BSc/MChem	3, 4				●		○	◐		●	◐	4
Southampton *www.soton.ac.uk*														
Chemistry		BSc/MChem	3, 4		○		●		○	◐	⊛	●	◐	3
Strathclyde *www.strath.ac.uk*														
Applied chemistry and chemical engineering	①	BSc/MSci	4, 5			◑	●		○	◐			◐	1
Chemistry	②	BSc/MSci	4, 5			◑	●		○	◐		●	◐	1
Chemistry with drug discovery		MSci	5				●					●		0
Forensic and analytical chemistry		MSci	5			◑	●		○	◐		●		0
Sunderland *www.sunderland.ac.uk*														
Chemical and pharmaceutical science		BSc	3, 4	●	○		●	◡		◐	⊛	●		0
Chemistry	①	BA/BSc	3, 4				●	◡	○	◐	⊛	●	◐	23
Surrey *www.surrey.ac.uk*														
Chemistry	①	BSc/MChem	3, 4, 5	●			●		○	◐		●	◐	2
Chemistry with forensic investigation		MChem	4							◐		●		0
Computer-aided chemistry		MChem	4	●					○	◐	⊛	●		0
Medicinal chemistry		BSc	4						○	◐		●		0
Sussex *www.sussex.ac.uk*														
Chemistry	①	BSc/MChem	3, 4		○		●		○			●	◐	1
Chemistry with forensic science		BSc/MChem	3, 4				●					●		0

(continued) **Table 2a** First-degree courses in **Chemistry**

Institution Course title (① ② ③ see combined subject list – Table 2b)	**Degree**	**Duration** Number of years	**Foundation year** ● at this institution	○ at franchised institution	◑ second-year entry	**Modes of study** ● full-time; ◗ part-time	○ time abroad	◐ sandwich	✪ **Modular scheme**	**Course type** ● specialised;	◐ combined	**No of combined courses**
Teesside *www.tees.ac.uk*												
Applied chemistry	BSc	3, 4	●			●	○	◐		●		0
Forensic chemistry	BSc	3, 4	●			●	○	◐		●		0
UCL *www.ucl.ac.uk*												
Chemical physics	BSc/MSci	3, 4				●				●		0
Chemistry ①	BSc/MSci	3, 4				●				●	◐	3
Medicinal chemistry	BSc/MSci	3, 4				●	○			●		0
Warwick *www.warwick.ac.uk*												
Biomedical chemistry	BSc	3, 4, 5		○		●				●		0
Chemistry ①	BSc/MChem	3, 4, 5		○		●	○	◐		●	◐	1
Chemistry with medicinal chemistry	BSc/MChem	3, 4, 5		○		●	○	◐		●		0
York *www.york.ac.uk*												
Chemistry ①	BSc/MChem	3, 4				●	○	◐	✪	●	◐	2
Chemistry, biological and medicinal chemistry	BSc/MChem	3, 4				●	○	◐	✪	●		0

Subjects available in combination with chemistry It is not possible to describe in the space available here the many different ways in which combined courses are organised, so you should read prospectuses carefully. For example, you should find out if the subjects are taught independently of each other or if they are integrated in any way. Combined courses in modular schemes often provide considerable flexibility, allowing you to vary the proportions of the subjects and include elements of other subjects. However, this means that you may lose some of the benefits of more integrated courses. TABLE 2b shows subjects that can make up between a third and a half of your programme of study in combination with chemistry in the courses listed in TABLE 2a. You can use TABLE 2b to find out where particular combinations are available. For example, if you are interested in combining chemistry with French, first look up French in TABLE 2b to find the institutions offering French in combination with chemistry. You can then use the index number given after the institution name to find which of the courses listed for the institution in TABLE 2a can be combined with French. If there is no index number after an institution's name in TABLE 2b, that is because there is only one course at that institution in the appropriate part of TABLE 2a.

Note that the names given in the table for the combined subjects have been standardised to make comparison and selection easier. This means that the name used at a particular institution may not be exactly the same as that given in the table: for example, 'psychological studies' may appear as 'psychology'. However, in nearly all cases it will be very similar, so you should not find much difficulty in identifying a particular course combination when you look at the prospectus.

Table 2b

Subjects to combine with **Chemistry**

Accountancy/accounting Aberdeen①
American studies Sunderland①, Sussex①
Archaeology Reading①
Art history Sunderland①
Astronomy Nottingham Trent①
Astrophysics Keele① ②
Biochemistry Heriot-Watt①, Huddersfield①, Keele① ②, Leeds①
Biological sciences Cardiff, London Metropolitan①
Biology Aston②, Glamorgan, Glasgow①, Keele① ②, Kingston①, Leeds①, Manchester Metropolitan①, Newcastle①, Surrey①
Biomedical sciences Northumbria①
Business computing Keele②
Business studies Aston②, Birmingham①, Greenwich①, Huddersfield①, Hull①, Keele① ②, Kingston①, Liverpool①, Manchester①, Sunderland①
Chemical engineering Aston①, Huddersfield①, Leeds①, Paisley①, Sheffield①, Strathclyde①
Computer science Aston②, Bangor, Belfast①, Heriot-Watt①, Leeds①, St Andrews①
Computing Kingston①, Manchester Metropolitan①, Sunderland①
Criminology Keele① ②
Dance Sunderland①
Economics Heriot-Watt①, Manchester Metropolitan①
Education Greenwich①, Manchester Metropolitan①, Strathclyde②, Sunderland①
English Keele① ②, Manchester Metropolitan①, Sunderland①
English language Aston②
Enterprise/entrepreneurship Greenwich①, Sheffield①
Environmental policy Manchester Metropolitan①
Environmental science Aston②, Birmingham①, Keele① ②
Environmental studies York①
European languages UCL①
European studies Aston②, Manchester Metropolitan①, Newcastle①, Sunderland①
Finance Keele① ②
Forensic science Glasgow①, Heriot-Watt①, Huddersfield①, Keele① ②, Leicester①, London Metropolitan①, Manchester Metropolitan①, Surrey①
French Aston②, Birmingham①, Greenwich①, Heriot-Watt①, Leeds①, St Andrews①
French studies Kingston①
Gender studies Sunderland①
Geography Keele① ②, Manchester Metropolitan①, Sunderland①
Geology Keele①, St Andrews①
German Heriot-Watt①, Leeds①, St Andrews①
Health studies Sunderland①
History Keele① ②, Manchester Metropolitan①, Sunderland①
History/philosophy of science Leeds①
Human biology Kingston①
Human geography Keele① ②
Human psychology Aston②
Human resource management Greenwich①, Keele① ②
Information systems Manchester Metropolitan①
Information technology Loughborough①, Sheffield①
International politics Manchester Metropolitan①
International studies/relations Aston②, Keele① ②
Internet technology Kingston①, St Andrews①
Journalism Sunderland①
Language studies Manchester Metropolitan①
Law Bristol①, Keele① ②
Linguistics Sunderland①
Management science Keele① ②, Manchester Metropolitan①
Management studies Aberdeen①, Bangor, Bath①, Edinburgh① ②, Heriot-Watt①, Imperial College London①, Leeds①, Leicester①, Newcastle①, Nottingham①, UCL①, Warwick①, York①
Marketing Greenwich①, Keele① ②, Manchester Metropolitan①
Materials science Heriot-Watt①
Mathematics Aberdeen①, Aston②, Glasgow①, Keele① ②, Leeds①, Manchester Metropolitan①, St Andrews①, Sheffield①, Southampton, UCL①
Media studies Sunderland①
Medical science Southampton
Modern languages Aberdeen①
Multimedia Manchester Metropolitan①
Music Keele① ②, Sunderland①
Music technology Keele① ②, Manchester Metropolitan①
Neuroscience Keele① ②
Nutrition London Metropolitan①
Oceanography Liverpool①, Southampton
Pharmaceutical sciences Brighton
Pharmacology Birmingham①, Kingston①, Leeds①, St Andrews①
Philosophy Keele① ②, Leeds①, Manchester Metropolitan①
Photography Sunderland①
Physical education Bangor
Physical geography Keele① ②
Physics Aberdeen①, Cardiff, Keele① ②, Leeds①, Nottingham①, St Andrews①

(continued) **Table 2b**

Subjects to combine with **Chemistry**

Subject	Institutions
Physiology	Sunderland①
Politics	Aston②, Keele① ②, Manchester Metropolitan①, Sunderland①
Psychology	Birmingham①, Keele① ②, Manchester Metropolitan①, Paisley①, Sunderland①
Public administration	Manchester Metropolitan①
Public policy	Aston②
Sociology	Keele① ②, Sunderland①
Spanish	Greenwich①, Heriot-Watt①, St Andrews①
Sports science	Bangor, London Metropolitan①, Loughborough①
TEFL	Sunderland①
Technology	Aston②
Tourism	Manchester Metropolitan①, Sunderland①

Other courses that may interest you The following courses are closely related to those in this Guide, so if you are interested in chemistry, you should perhaps consider these as well. They include courses where you spend less than half your time studying chemistry, as well as courses providing a more intensive study of a specialised aspect of chemistry.

- Chemistry for offshore industry (Aberdeen)
- Marine chemistry (Bangor)
- Biomedical materials chemistry (Aberdeen).

The following Guides also contain groups of courses that may interest you:

- *Biological Sciences*
- *Engineering: Chemical Engineering*
- *Medical and Related Professions* for courses in pharmacology and biomedical sciences
- *Agricultural Sciences and Food Science and Technology*: courses in food science have a large component of chemistry.

Basic content The foundations of chemistry are well established, so, apart from specialised courses dealing with one particular aspect of the subject, the basic content of degree courses is similar at all institutions. Even the specialised courses have a great deal of material in common with other courses, though the emphasis and perspective may differ.

The courses usually build on the work done at school, but in addition to the familiar areas of inorganic, organic and physical chemistry, you will probably come across other areas of the subject, such as theoretical chemistry (which includes the use of quantum mechanics to describe molecular structure) and analytical chemistry. However, none of the boundaries are very sharp, and as time goes on they are being progressively blurred. In some areas, such as organometallic chemistry, the traditional divisions have disappeared completely.

Many courses contain a significant amount of computer-based work. You are probably already familiar with using word-processing software, spreadsheets, databases and the internet, but support will be available if you need it. You will be taught how to use a variety of specialist packages aimed at visualising molecular geometry and carrying out simple molecular modelling. You will also need to become skilled in using a chemical drawing package.

Final-year specialisation In the final year of a BSc degree or final two years of an MChem/MSci degree you will probably be given the opportunity to choose modules with a specialised content. These often deal with the frontiers of chemistry, which means they need to be regularly updated and can be very exciting. Options allow you to follow a particular interest to a greater depth and can often add considerably to the enjoyment and benefit you gain from the course.

The range of topics available varies widely from institution to institution, as does the amount of choice you have. TABLE 3a shows what percentage of your time in the final year is spent on compulsory topics in chemistry, and what percentage can be spent on optional topics (other topics outside chemistry make up the balance). It also shows which topics are on offer on each course, and whether they are compulsory components ●, optional components ○, or with some compulsory and additional optional components ◑.

Table 3a

Course content

Institution Course title ① ②: see notes after table ○ optional; ● compulsory; ◑ compulsory + options	**Final year chemistry content** Compulsory (%)	Optional (%)	Analytical chemistry	Organic synthesis	Organometallic chemistry	Catalysis	Materials chemistry	Polymers	Solid-state chemistry	Theoretical chemistry	Photochemistry	Chemometrics	Mass spectrometry	Crystallography	Computational chemistry	Macromolecules	Biotechnology	Biological chemistry	Bioinorganic chemistry	Pharmaceutical/pharmacological chemistry	Medicinal chemistry	Food chemistry	Environmental chemistry
Aberdeen																							
Chemistry ①	60	40	○	◑	●	●	◑	◑	○	●	◑		●	◑	○	◑		○	○	○	○		○
Medicinal chemistry ②	60	40	○	●	◑	●		◑	○	◑	○		●	●	○	◑	●	●	○	●	●		○
Aston																							
Applied chemistry ①	75	25	●	●	●	●	●	◑		●	●		●	◑	●	●	●			◑			○
Chemistry ②	75	25	●	●	●	●	●	◑		●	●		●	◑	●	●	●			◑			○
Chemistry (biotechnology)	75	25	●	●	●	●	●	◑		●	●		●	◑	●	●	●	◑		◑			○
Chemistry (environmental management)	75	25	●	●	●	●	●	◑		●	●		●	◑	●	●	●	◑		◑			○
Chemistry (management studies)	75	25	●	●	●	●	●	◑		●	●		●	◑	●	●	●	◑		◑			○
Bangor																							
Chemistry	100			●	●	●	●	●	●	●	●		●	●	●	●	●	○	●				○
Bath																							
Chemistry ①	60	40	●	●	●	○	○	●	●	●	●			●	○	○		○	○	○	○		●
Chemistry for drug discovery	60	40	●	●	●	◑	○	●	●	●	●			●	○	○		●	○	●	●		○
Belfast																							
Chemistry ①	80	20	○	◑	●	◑		○	●					●		○		○	○	○	○		○
Birmingham																							
Chemistry	50	17	◑	◑	◑	◑	●	●	◑	○	●		◑	◑	○	○	○	◑	●	○	◑		◑
Bradford																							
Chemistry with pharmaceutical and forensic science ①	8	40	◑	◑	●	○	◑	○		○	○		●	●		●		◑		●			
Brighton																							
Pharmaceutical and chemical science			●	●	●	●	○	○	○	●	○	●	●	○	○	●	○	●	○	●	○	○	○
Bristol																							
Chemical physics	50	0					●	●	●	●	●			●	●	●							
Chemistry	100	0	●	●	●	●	●	●	●	●	●	●	●	●	●	●		●	○		●		●
Bristol UWE																							
Biological and medicinal chemistry	15	85	●	●	○	○	○	○	○				●	○		●	○	●	○	●	●		
Cambridge																							
Natural sciences (chemistry)	25	75		◑	◑	◑	◑	◑	◑	◑				◑				◑	◑		◑		
Cardiff																							
Chemistry ①	83	17	◑	●	●	●	◑	◑	●	◑	◑			●	◑	○	○	○	○	○	○		○
Dundee																							
Pharmaceutical chemistry	50	25	●	●									●	●	●	●				●			
Durham																							
Chemistry	33	67	○	◑	◑	◑	○	◑	○	○	◑	○	○	○	○	○	○	○	○	○	○	○	○
East Anglia																							
Biological and medicinal chemistry	40	60		○	○													●	○	●	●		
Chemical physics	50	50		○			○	○		●													
Chemistry ①	0	100	○	○	○	○	○	○	○	○	○		○					○	○	○	○		○
Chemistry with analytical and forensic science ②	75	25	●	●	○	○	○	○	○	○	○		○					○	○	○	○		
Chemistry with analytical science ③	40	60	●	○	○	○	○	○		○	○		●	○		○		○	○	○	○		●
Pharmaceutical chemistry	40	60	●	○			○	○										○		●	●		
Edinburgh																							
Chemistry ①	75	25		●	○	◑	◑	●	○	●	○	◑	●	◑	◑	●	○	●	●	●	○	◑	◑
Chemistry with environmental chemistry ②	75	25	●	●	○	◑	◑	●	○	●	○	◑	●	◑	◑	●	○	●	●	○	○	◑	●

(continued) **Table 3a**

Course content

Institution Course title ① ②: see notes after table ○ optional; ● compulsory; ◑ compulsory + options	**Final year chemistry content** Compulsory (%)	Optional (%)	Analytical chemistry	Organic synthesis	Organometallic chemistry	Catalysis	Materials chemistry	Polymers	Solid-state chemistry	Theoretical chemistry	Photochemistry	Chemometrics	Mass spectrometry	Crystallography	Computational chemistry	Macromolecules	Biotechnology	Biological chemistry	Bioinorganic chemistry	Pharmaceutical/pharmacological chemistry	Medicinal chemistry	Food chemistry	Environmental chemistry
Edinburgh (continued)																							
Medicinal and biological chemistry ③	75	25	●	●	○	○	◑	◑	○	●	○	◑	●	◑	○	●	●	●	●	●	●	○	○
Exeter																							
Biological and medicinal chemistry ①	50			●														●	●	●	●		
Glamorgan																							
Chemistry	80	20	●	●	●	●		●		●	●		●		●	●		●	●	●			
Glasgow																							
Chemistry	50	50	●	◑	●	◑	◑	◑	○	●	●			◑	○	●		◑	●	○			◑
Chemistry with medicinal chemistry ①	50	50	●	◑	◑	○	●		◑	◑	●		●	●		◑		●	●	●	●		
Greenwich																							
Analytical chemistry	100		●			●		○		○	○	●	●		○	○							○
Chemistry	25	75	○	○	○	○	○	○	○	○	○	○	○	○	○	○		○	○	○	○		○
Pharmaceutical chemistry ①	75	25	●	●	○	○				○	○	●	○	○	○					●	●		
Heriot-Watt																							
Chemistry	80	20	●	●	●	●	●	●	●	●	●		●	●		●		○	○	○	○		●
Chemistry with pharmaceutical chemistry	80	20		●	●	●				●	●						●	●	●	●	●	●	
Huddersfield																							
Chemistry	75	25	◑	●	●	●	●	○	●	●			●			◑	○	○			○	○	○
Chemistry with analytical chemistry	75	25	●	●	●	●	●		●	●			●					○				○	○
Chemistry with medicinal chemistry	75	25	◑	●	●	●	●		●	●			●			◑		●			●		
Hull																							
Chemistry ①	70	30	○	◑	○	○	○	○		○	◑	○			○	○		○	○	○	○		○
Chemistry with analytical chemistry and toxicology ②	100	0	●									●										●	●
Chemistry with forensic science and toxicology	100		●									●											●
Imperial College London																							
Chemistry		100	○	○	○	○	○	○	○	○	○		○	○	○	○		○	○	○	○		○
Chemistry with conservation science		100	○	○	○	○	○	○	○	○	○		○	○	○	○		○		○	○		○
Chemistry with fine chemicals processing		100	○	○	○	○	○	○	○	○	○		○	○	○	○							○
Chemistry with medicinal chemistry	80	20		●	●	○	○	○	○	○	○		○	○	○	○		○	○	●	●		○
Keele																							
Chemistry	50	25	◑	◑	◑	◑	◑	◑	◑	◑	○		◑	◑	○	○		◑	◑	◑	◑		
Medicinal chemistry ①	50	25	◑	◑		◑					○		◑	◑		◑		◑	○	◑	◑		
Kingston																							
Chemistry	50	75	●	●	●	○	○	○	○				●			○					○		●
Chemistry (applied)	50	75	●	●	●	○	●	●					●			○					●		●
Medicinal chemistry	30	60	●	◑											○		○	●		●	●		
Leeds																							
Applied chemistry ①	75	25	●	●	●	●	●	●			●		●		●	●				●	◑	◑	●
Chemistry	30	70	◑	◑	◑	◑	○	◑	◑	○	◑		◑	◑	○	○		○	○	○	○		○
Chemistry with analytical chemistry	75	25	◑	◑	◑	◑		◑	◑	◑	◑		◑	◑	◑			◑	◑		◑		◑
Colour and polymer chemistry ②	75	25	●	●	●		●	●			●		●		●	●							●
Medicinal chemistry	0	100	◑	◑	●	○		○		○			◑	◑	○			○	○	●	●		

(continued) Table 3a

Institution Course title ① ②: see notes after table ○ optional; ● compulsory; ◑ compulsory + options	Final year chemistry content Compulsory (%)	Optional (%)	Analytical chemistry	Organic synthesis	Organometallic chemistry	Catalysis	Materials chemistry	Polymers	Solid-state chemistry	Theoretical chemistry	Photochemistry	Chemometrics	Mass spectrometry	Crystallography	Computational chemistry	Macromolecules	Biotechnology	Biological chemistry	Bioinorganic chemistry	Pharmaceutical/pharmacological chemistry	Medicinal chemistry	Food chemistry	Environmental chemistry
Leicester																							
Chemistry ①	50	50	○	◑	◑	○	◑	○	◑	◑	○		●	◑	◑	○		○	○		○		
Pharmaceutical chemistry	50	50	○	◑	◑	○	◑	○	◑	◑	○		●	◑	◑	○		○	○	◑	○		
Liverpool																							
Chemistry ①	60	25		●	●	●		●	●				●					○		○	○		
Medicinal chemistry	75	25		●	●	●												●		●	●		
Medicinal chemistry with pharmacology ②	50	50		●	●	●	●										●	●		●	●		
Liverpool John Moores																							
Medicinal and analytical chemistry	50	20	●	●	●	●	○	○	○	●	○	○	●	○	●	○	●	●	○	●	●		
Medicinal chemistry	50	20	●	●	●	●	○	○	○	●	○	○	●	○	●	○	●	●	○	●	●		
Pharmaceutical science and biological chemistry	50	20	◑					○	◑	●		●	◑		◑		○	●		○	●		○
London Metropolitan																							
Biological and medicinal chemistry ①	60	20	○	●	○					○			○	○		○		●	○		●		
Chemistry	60	40	○	●	●					●			○	○		○		○			○		○
Loughborough																							
Chemistry ①	66	33	◑	◑	○	◑	●	○	●	●	○	◑	●	●	◑	◑	○	○	●	○	○	○	○
Chemistry with analytical chemistry ②	83	17	●	◑	●	●	○	●	●	●	○	●	●	●	◑	◑	○	○	●	●	○	●	●
Chemistry with forensic analysis	85	15	◑	◑	●	●	○	●	●	●	○	●	●	●	◑	◑	○	○	●	●	○	●	●
Medicinal and pharmaceutical chemistry ③	70	30	◑	◑	○	●	●	○	●	●	○	◑	●	●	◑	◑	○	●	●	●	●		○
Manchester Metropolitan																							
Chemical science ①	75	25	○	●	●	●		○	○		○					○	○	○	●		○		○
Chemistry ②	75	25	○	●	●	●		○	○		○		○			○		○	●		○		○
Medicinal and biological chemistry	75	25	○	●													●	●	●	●	●		
Newcastle																							
Chemistry ①	70	30		●	●	○	○	◑	●	◑	○		●	●	◑	●		○	●		○	◑	●
Chemistry with medicinal chemistry ②	70	30		●	●	○	○			◑	○		●	●	◑	●		●	●	●	●		
Northumbria																							
Applied chemistry	50	25	●	●	●	●	○	○	●	●	●	○	●		○	○							
Chemistry				●		●		●		●		●											
Nottingham																							
Chemistry	80	20	○	●	●	◑	○	◑	◑	◑	◑			◑	◑	◑	◑	◑	●	○	○		○
Nottingham Trent																							
Chemistry	50	50	●	●	●	●	○	○	●	○	●	○	●	●	○	●		○	●	○	○		○
Oxford																							
Chemistry			●	●	●	●	●	●	●	●	●		●	●	●	●		●	●	○	●		
Paisley																							
Chemistry	100		●	●	●			●	●	●	●	●	●	●		○			●		●		●
Plymouth																							
Chemistry (analytical) ①	83	17	●	●	●	●	○	○		●	●	○	●			○	○	○					○
Chemistry (applied) ②	100		●	●	●		●	●		●	●		●					●					
Reading																							
Chemistry	70	30		●	●	○	○	○	●	●	○			○	○	○			○	●	●		○

Chemistry

(continued) Table 3a Course content

○ optional; ● compulsory; ◑ compulsory + options

Institution Course title ① ②: see notes after table	**Final year chemistry content** Compulsory (%)	Optional (%)	Analytical chemistry	Organic synthesis	Organometallic chemistry	Catalysis	Materials chemistry	Polymers	Solid-state chemistry	Theoretical chemistry	Photochemistry	Chemometrics	Mass spectrometry	Crystallography	Computational chemistry	Macromolecules	Biotechnology	Biological chemistry	Bioinorganic chemistry	Pharmaceutical/pharmacological chemistry	Medicinal chemistry	Food chemistry	Environmental chemistry
St Andrews																							
Chemical sciences	70	30	●	●	○	○	○	○	○	●			●	○	○	○	○	○	○	○	○	○	●
Chemistry	35	25	●	●	◑	○	○	○	◑	●			●	○	○	○		○	○	○	○		●
Chemistry with catalysis	45	15	●	●	◑	○	○	○	◑	●			●	○	○	○		○	○	○	○		○
Chemistry with materials chemistry	45	15	●	●	◑	○	●	●	●	○			●	○	○	○		○	○		○		●
Chemistry with medicinal chemistry	35	25	●	●	◑	○	○	○	○	●			●	○	○	○		●	●	●	●		●
Sheffield																							
Chemical physics	50	0	◑	◑			◑	◑	◑							◑							
Chemistry ①	100	1		●	●	●	●	●	●	●	●		●	●	●			●	●				
Southampton																							
Chemistry ①	0	100	○	○	○	○	○	○	○	○	○				○				○	○	○		○
Strathclyde																							
Applied chemistry and chemical engineering	50	50	◑	◑	◑	◑	◑	◑	◑	●	◑		◑	●	●	◑					○		○
Chemistry	50	50	◑	◑	◑	◑	◑	◑	◑	●	◑		◑	●	●	◑					○		○
Forensic and analytical chemistry ①	50	50	●	◑	◑	◑	◑	◑	◑	●	◑		◑	●	●	◑					○		○
Sunderland																							
Chemical and pharmaceutical science ①	75	25	◑	●	●	●				◑	○		●		○	○	●			●	○	○	○
Surrey																							
Chemistry	30	40	○	◑	◑	○	○	◑	◑	○		○	●	○	○	○			○	○	○		○
Computer-aided chemistry	40	40	○	◑	◑		○	◑	◑	○		●		○	●	◑							○
Medicinal chemistry	30	40	◑	●	◑	◑	◑	◑	◑	○		○	○	○	○	◑		◑	◑	●	●		
Sussex																							
Chemistry ①	50	25	●	○	○	○		○	○	○	○				○	○			○	○	○		○
Teesside																							
Applied chemistry	50	50	●	●	●	●	●	○	◑	○		●				○		●	●		●		
Forensic chemistry	15	15	●	●	●	○	○	○			●	●	●			●			●				
UCL																							
Chemical physics	38	50		◑	○	○	○	○	◑	◑	◑		○	○	◑	○		○	○		○		
Chemistry	40	60		◑	◑	○	○	○	◑	◑	◑		○	○	○	○		○	○		○		
Medicinal chemistry ①	60	10		◑	○	○	○	○	○	○	◑		○	○	○	○	○	○	●	●	●		
Warwick																							
Biomedical chemistry	25	55	◑	◑	◑	○		○			○		●		○	○	○	◑	●	○	◑		
Chemistry	45	55	●	◑	◑	◑	○	○	◑	◑	○	○	○	○	○	○		○			○		○
Chemistry with medicinal chemistry	45	55	●	●	◑	◑	○	○	◑	◑	○	○	○	○	○	○		●	●	●	●		○
York																							
Chemistry ①	60	40	◑	●	◑	◑	◑	◑	◑	◑	○		◑	●	○	◑	○	◑	◑	◑	◑		○

Aberdeen ① Organic spectroscopy ●; biosynthesis ○; physical organic ○; thermodynamics ●; electrochemistry ○; surface chemistry ○
② Biomedical science

Aston ① ② Physical inorganic chemistry; surface analysis

Bath ① Electrochemistry ○; neutron scattering ○; physical organic chemistry ●; spectroscopy ●; statistical thermodynamics ●; coordination chemistry ●

Belfast ① Thermodynamics and kinetics ●; spectroscopy ●; industrial chemistry ○

Bradford ① Forensic anthropology ●; forensic biochemistry ●; forensic archaeology ○; drug analysis ○; interpretation of forensic evidence ○

Cardiff ① Molecular dynamics ●; surface and molecular spectroscopy ○

East Anglia ① Spectroscopy ○ ② Forensic science ●; spectroscopy ● ③ Spectroscopy ○

Edinburgh ① Main-group chemistry ●; clusters ●; industrial chemistry ○; electrochemistry ●;

colloids ○; reaction dynamics ● ②Atmospheric and hydrospheric chemistry; environmental toxicology management and modelling; environmental geochemistry ③Biophysical chemistry ●
Exeter ①Physical chemistry for the life sciences; forensic science
Glasgow ①Pharmacology ●; industrial medicinal chemistry ●; stereochemistry ●
Greenwich ①Technology transfer ○
Hull ①Kinetics; atmospheric chemistry; combustion; structure determination; electrochemistry; surface chemistry; liquid crystals ②Toxicology; molecular analytical spectroscopy; separation science; electrochemical analysis
Keele ①Modern drug design; antibiotics and chemotherapy
Leeds ①Surface and colloid chemistry ●; colour chemistry ●; imaging science ● ②Colour chemistry ●; surface coatings ●
Leicester ①Molecular spectroscopy ◑; electrochemistry ○
Liverpool ①Electrochemistry ●; kinetics ●; colloids ●; sustainable chemistry ○; atmospheric chemistry ○; applications of group theory ○ ②Biocatalysis ○; protein structure and dynamics ○; natural product chemistry ○; nucleic acid and peptide chemistry ○
London Metropolitan ①Natural product chemistry ○
Loughborough ① ②Chemotherapy of infection ○ ③Molecular pharmacology ●; pharmacokinetics and drug metabolism ●
Manchester Metropolitan ① ②Frontiers in chemistry ●
Newcastle ①Reaction mechanism ●; main group chemistry ●; energetics ○; surface science ○ ②Reaction mechanism ●; main group chemistry ●; energetics ○; surface science ○; toxicology ●
Plymouth ①Forensic analysis ● ②Chemical engineering
Sheffield ①MChem: statistical mechanics and dynamic processes ●; many options
Southampton ①Kinetics ○; surface sciences ○; liquid crystals ○; electrochemistry ○; transition metals ○; laser spectroscopy ○; supramolecular chemistry ○
Strathclyde ①Forensic chemistry ●
Sunderland ①Forensic analysis
Sussex ①Molecular dynamics; modern materials
UCL ①Pharmacology ●
York ①Mechanistic inorganic and organic chemistry ◑; NMR spectroscopy ●; statistical thermodynamics ◑; advanced spectroscopy ○; industrial management ○; fast reactions and short-lived intermediates ○

MChem/MSci courses TABLE 2a lists many courses that can lead to the award of either a BSc (occasionally BA) or an MChem/MSci degree. BSc degrees usually involve studying for three years. They provide an excellent training in the chemical sciences and can open doors to careers in a large range of employment sectors. MChem/MSci degrees usually last for four years and provide a more in-depth study of chemistry than BSc degrees. They normally include a significant research project and offer more opportunity to develop skills such as presentation, communication and problem-solving.

Most of the tables in this Guide give information specifically for the BSc stream (the tables show information for the MChem/MSci course where there is no BSc). Where both BSc and MChem/MSci are available, much of the information will also apply to the MChem/MSci streams, but TABLE 3b shows you where there are differences for the MChem/MSci stream.

Table 3b Differences between MChem/MSci and BSc courses

Institution	Course title	Differences
Aberdeen	Chemistry	Industrial placements; additional project work; specialist lecture; advanced practical techniques; self-directed study leading to essay and oral presentations
Aston	*Both courses*	Significantly enhanced practical content; 'mini' thesis
Bangor	Chemistry	Higher entry grades required; additional research project; advanced topics in chemistry; additional management topics
Bath	Chemistry	Year 3 options: (1) assessed industrial training, with chemistry units and Professional Studies by distance learning (2) study year abroad in USA or Europe, with chemistry units by distance learning (3) full-time study at Bath. Year 4: more advanced options and 2-semester research project
Belfast	Chemistry	Year 3: set practical work; 1/3 of year 4 on project
Birmingham	Chemistry	Different laboratory work in final year; more extensive research project
Bradford	Chemistry with pharmaceutical and forensic science	BBB A-level grades required; research skills; extensive final-year research project; communication and group work skills; selected chemistry topics in greater depth
Bristol	*Both courses*	Year 3 practical component includes advanced practical topics; year 4 research project requires 22–24 hours/week
Cambridge	Natural sciences (chemistry)	Project/dissertation in final year (25% of marks); no compulsory coursework: free choice from wide range of options
Cardiff	Chemistry	Higher entry points (300–340); research project one-third of final year; more work in small groups; separate course from year 3
Durham	Chemistry	50% of work in final year is research project; training in research methods in areas including industrial chemistry, health and safety and analytical chemistry; choice of advanced lectures complementing project
East Anglia	Biological and medicinal chemistry	Year 3 can be spent in N America or industry; extended project occupies 50% of final year
	Chemical physics	Year 3 can be spent in Europe, N America or industry; extended project occupies 50% of final year
	Chemistry	Year 3 can be spent in Europe, N America or industry; extended project occupies 50% of final year
Edinburgh	Chemistry	Advanced chemistry topics; extended research project; further transferable skills development; year abroad or in industry
	Chemistry with environmental chemistry	Masters level courses in environmental chemistry; optional year abroad or in industry
	Medicinal and biological chemistry	Specialist courses at Masters level in final years; extended research project work
Glasgow	Chemistry	Enhanced research project, more advanced training in research methods; additional coursework in years 3 and 4; summer vacation written exercise
	Chemistry with medicinal chemistry	Literature review between levels 3 and 4; longer research project, more advanced training in research methods; additional subject coverage; more in-depth coverage of some coursework; MSci placements in industry/Europe
Greenwich	Chemistry	Longer project; case studies; more mathematics
Heriot-Watt	*Both courses*	Broader scope including management skills and chemical engineering; large research project; communication skills
Huddersfield	Chemistry	Extra breadth and depth; longer research project
Hull	Chemistry	Wider choice; more advanced modules; extended research project
	Chemistry with analytical chemistry and toxicology	Wider choice; more advanced modules; extended research project
Imperial College London	Chemistry	More advanced topics
Keele	Chemistry	Advanced option modules; greater breadth and depth; half year 4 on research project and dissertation; group-based experimental design; health and safety; transferable skills
Kingston	*Both courses*	More courses and project work
Leeds	*Both courses*	Practical project is 50% of final year

(continued) **Table 3b** Differences between MChem/MSci and BSc courses

Institution	**Course title**	**Differences**
Leicester	Chemistry	Final 2 years strongly research-oriented; final-year research project has more than twice the laboratory time of BSc research projects
	Pharmaceutical chemistry	Final 2 years strongly research-oriented; final-year research project has more than twice the laboratory time of BSc research projects
Liverpool	Chemistry	After 18 months, MChem courses become more detailed. MChem year 4 project 38% of year. Year 4 options include biocatalysis, protein structure and dynamics, natural product chemistry, chemical nanotechnology
Loughborough	Chemistry	All topics become compulsory; research techniques, experimental design and communication skills extended; major research project
	Chemistry with analytical chemistry	All topics become compulsory; research techniques, experimental design and communication skills extended; major research project
	Chemistry with forensic analysis	Greater proportion of compulsory modules; research techniques, experimental design and communication skills; major extended research project
	Medicinal and pharmaceutical chemistry	All topics become compulsory; research techniques, experimental design and communication skills extended; major research project
Manchester Metropolitan	Chemistry	Professional skills; year-long research project; further chemistry topics; extended practicals
	Medicinal and biological chemistry	Professional skills; year-long research project; specialised biological and medicinal options
Newcastle	*Both courses*	Extended research project and advanced topics in year 4
Northumbria	Applied chemistry	Greater depth; major research project
Nottingham	Chemistry	Optional year in Europe, USA or Australia, or with 5-year option, year in industrial laboratory
Nottingham Trent	Chemistry	Extended project; research methodology; additional chemistry topics; advanced techniques; transferable skills
Reading	Chemistry	Major project (50% of time)
St Andrews	*Both courses*	Industrial placement year; advanced problem-solving and communication/transferable skills emphasised
Sheffield	Chemical physics	Project contributes 50% of final-year mark
	Chemistry	Recent advance topics in chemistry; research project 50% final-year mark
Southampton	Chemistry	Full-time 6-month period of industrial experience or academic research; fully accredited by RSC (BSc associate membership only)
Strathclyde	*Both courses*	One-year industrial placement
Surrey	Chemistry	Minimum average mark of 55% in first 2 years required; year 3 is substantial research project in industry; additional modules on research techniques in final year
Sussex	Chemistry	Additional options
UCL	Chemical physics	Year 4: 50% research project
	Chemistry	50% of final year spent on research project
	Medicinal chemistry	50% of final year spent on research project
Warwick	*Both courses*	Project worth 50% of final year; 6-month placement
York	Chemistry	Year 3: advanced training module instead of research project; year 4 options: (1) in York (additional chemistry content, project planning, literature review, research project) (2) university in France, Germany or Italy, including project work (3) programme of structured and assessed industrial training, including project work

Supporting and subsidiary content In all courses you will be required to study at least one other subject in addition to chemistry for one or two years. Some of these subjects, such as mathematics, are taught to support your work in chemistry; others, such as a foreign language, are designed to teach you additional skills, or may be there just to broaden your horizons. The subjects offered, the time they occupy and whether they are compulsory or optional varies considerably from institution to institution. For full information consult prospectuses or institutions directly. Institutions that run modular degree schemes can be particularly flexible with the topics they offer: see TABLE 2a for courses that are part of modular degree schemes in which you are relatively free to choose topics from a large number of different subjects.

TABLE 3c gives an indication of the pattern for courses listed in this Guide. In each column, the position of the symbols indicates whether the subject is available in the first, intermediate or final years.

Table 3c Supporting and subsidiary content in first/intermediate/final years

● compulsory; ○ optional; ◑ compulsory + options

Institution Course title	Mathematics/statistics	Physics	Other physical sciences	Biological science	Computing for chemists	Computer science	History of science	Management	Other business	Foreign language	Social science	Humanities	Communications	Personal skills
Aberdeen														
Chemistry	○○	○○○	○○	○○	●	○○	○○	○○○	○○○	○○○	○	○	○●●	○●●
Medicinal chemistry	○	○	○	●●◑	●	○		○	○	○	○	○	○	●●●
Aston														
Applied chemistry	●		●	●		●		●	●	○○			●	●
Chemistry	●		●	●		●		●	●	○○			●	●
Chemistry (biotechnology)	●		●	●		●		●	●	○			●	●
Chemistry (environmental management)	●		●	●		●		●	●	○○			●	●
Chemistry (management studies)	●		●	●		●		●	●	○○	○		●	●
Bangor														
Chemistry	○○	○○	○○	○○	○●	○○		○○	○○	○○	○○	○○	○○	○○
Bath														
Chemistry	●○	○○	○○	○○	○	●		○○○		○○○	○	○		
Chemistry for drug discovery	●○	○○	○○	●●	○	●				○○○	○	○		
Belfast														
Chemistry	○●	○	○	○		○	○			○	○	○		
Birmingham														
Chemistry	◑○		○	○	●●●				○	○○			●●	●●
Bradford														
Chemistry with pharmaceutical and forensic science	●●		●◑◑	●○○	●○								●○	●○
Brighton														
Pharmaceutical and chemical science	●●○	○○○	○○○	●●●	○○○	○○○	○○○	○○○	○○○	○○○	○○○	○○○	○○○	●●●
Bristol														
Chemical physics	●●○	●●●			●	●●							●●	●●
Chemistry	○	○	○	○	●	○				○○			●●	●●
Bristol UWE														
Biological and medicinal chemistry	●●			●●●	●●○	●●○				○○			○○	○○
Cambridge														
Natural sciences (chemistry)	○○	○○	○○	○○	○○	○	○			○			●	

Supporting and subsidiary content in first/intermediate/final years

Institution Course title ● compulsory; ○ optional; ◑ compulsory + options	Mathematics/statistics	Physics	Other physical sciences	Biological science	Computing for chemists	Computer science	History of science	Management	Other business	Foreign language	Social science	Humanities	Communications	Personal skills
Cardiff														
Chemistry	○	○	○	○						○○			●●	●●
Dundee														
Pharmaceutical chemistry			○	●●	●	○				○	○	○	●●	●●●
Durham														
Chemistry	●○○	○○○	○○○	○○○	●○○	○○○	○○○			○○○			○●●	●●●
East Anglia														
Biological and medicinal chemistry	○	○		●●●	○					○				○
Chemical physics	●●	●	●											
Chemistry	○○	○○○	○	○○○		○	○	○○○		○		○○○		○○
Chemistry with analytical and forensic science	○		○											
Chemistry with analytical science	○○	○○	○○	○○		○○				○	○	○		
Pharmaceutical chemistry	○		○											
Edinburgh														
Chemistry	●○	○○	○○	○○	○●	○○	○○	○○○	○○	○○	○○	○○	●	●
Chemistry with environmental chemistry	●○	○○	○○	○○	○●	○○	○○	○○○	○○	○○	○○	○○	●	●
Medicinal and biological chemistry	●○	○○	○○	●●	○●	○○	○○	○○	○○	○○	○○	○○	●	●
Exeter														
Biological and medicinal chemistry	●			●●●	●	●							●	●●●
Glamorgan														
Chemistry	●○○	●○				●		○○○					●	○○○
Glasgow														
Chemistry	○○	○○	○○	○○	●●	○○		○○		○○	○○	○○	◑◑●	◑◑●
Chemistry with medicinal chemistry	○○	○○	○○	○○●	●	○○				○○	○○	○○	◑◑●	◑◑●
Greenwich														
Analytical chemistry	●●●			●	●								●●●	●●●
Chemistry	●●○			○	●			○○○	○○○	○○○			●●●	●●●
Pharmaceutical chemistry	●●●			●	●								●	●
Heriot-Watt														
Chemistry	●○	○○	○	○○	●○	○○	○○	○○●	○○	○○			●●●	●●●
Chemistry with pharmaceutical chemistry	●			●●	●			●					●●●	●●●
Huddersfield														
Chemistry	●			○○	●●				○○					●
Chemistry with analytical chemistry	●			○	●									●
Chemistry with medicinal chemistry	●			●	●									●
Hull														
Chemistry	○	○○		○○		○○○		○	○○	○			● ●	● ●
Chemistry with analytical chemistry and toxicology	○	○○		○○		○○○							● ●	● ●
Chemistry with forensic science and toxicology	○	○○		○○		○○○							● ●	● ●
Imperial College London														
Chemistry	● ○	◑	○ ○	○		●	○	○		◑ ○		◑ ○	●	●◑

Chemistry

(continued) Table 3c

Supporting and subsidiary content in first/intermediate/final years

● compulsory; ○ optional; ◑ compulsory + options

Institution / Course title	Mathematics/statistics	Physics	Other physical sciences	Biological science	Computing for chemists	Computer science	History of science	Management	Other business	Foreign language	Social science	Humanities	Communications	Personal skills
Imperial College London (continued)														
Chemistry with conservation science	●○	○		○		●	○○			◑		◑○	●	●●
Chemistry with fine chemicals processing	●○	◑	○○			●	○	○		○		○○	●	●
Chemistry with medicinal chemistry	●			●		●						◑	●	●●
Keele														
Both courses	◑							○	○	○	○	○	●●●	●●●
Kingston														
Chemistry	●		○		●					○			●	●
Chemistry (applied)	●		○		●					○			●	●
Medicinal chemistry	●			●	●					○			●	●
Leeds														
Applied chemistry	○	○	○		○○	●●		○		○○			◑◑●	●●●
Chemistry	○	○		○		○	○	○		○				○
Chemistry with analytical chemistry	○	○		○		○	○	○		○				○
Colour and polymer chemistry	○				○○	●●		○		○○			◑	●●●
Medicinal chemistry	○	○		○	○	○	○	○		○				○
Leicester														
Chemistry	●				●	●		○		○			●	●
Pharmaceutical chemistry	●				●	●		○		○			●	●
Liverpool														
Chemistry	◑◑	◑◑	○○	○○	●●●			○○					●●	●●●
Medicinal chemistry	◑◑	◑◑		●●	●●●									●●●
Medicinal chemistry with pharmacology	◑◑	◑◑	○○	●●	●●●									
Liverpool John Moores														
Medicinal and analytical chemistry	●				●								●	●
Medicinal chemistry	●				●								●	●
Pharmaceutical science and biological chemistry	○○	○○	○○	●	○○	○○	○	○		○○	○○	○○	○○	●
London Metropolitan														
Biological and medicinal chemistry	●○○	○○○	○○○	●●●		○○○	○○○	○○○	○○○	○○○	○○○	○○○	○○○	○○○
Chemistry	●○○	○○○	○○○	○○○		○○○	○○○	○○○	○○○	○○○	○○○	○○○	○○○	○○○
Loughborough														
Chemistry	●○	●	●●◑	○●○	●	●							●	●
Chemistry with analytical chemistry	●○	●	●●◑	○●○	●	●							●	●
Chemistry with forensic analysis	●○	●	●●◑	○●○	●	●							●	●
Medicinal and pharmaceutical chemistry	●○	●	●●◑	○●●	●	●							●	●
Manchester Metropolitan														
Chemical science	●○○		○○○	○○○	●	○○○				○○○			●	●
Chemistry	●○○		○○○	○○○	●○					○○○			●	●
Medicinal and biological chemistry	●			●●●	●○					○			●	●

(continued) Table 3c

Supporting and subsidiary content in first/intermediate/final years

● compulsory; ○ optional; ◑ compulsory + options

Institution Course title	Mathematics/statistics	Physics	Other physical sciences	Biological science	Computing for chemists	Computer science	History of science	Management	Other business	Foreign language	Social science	Humanities	Communications	Personal skills
Newcastle														
Chemistry	○	○		○		○		◑◑◑		◑◑			○	●●●
Chemistry with medicinal chemistry	○	○		○		○							○	●●●
Northumbria														
Applied chemistry	●				●			●		◑◑◑				
Nottingham														
Chemistry	◑○○	○○○	○○○	○○○		○○○	○○○	○○○	○○○	○○○	○○○	○○○		
Nottingham Trent														
Chemistry	○○	○○		○	○○	○○		○○	○	○○			●●●	●●●
Oxford														
Chemistry	●		○		●		○			○				
Paisley														
Chemistry	○	○	●	○		●							● ●	● ●
Plymouth														
Chemistry (analytical)	◑		○	○	●	○			○	○			●●●	●●●
Chemistry (applied)	◑		○	○	◑	○			○	○			●●●	●●●
Reading														
Chemistry	○	○		○	○○	○		○		○○		○	●●	○○
St Andrews														
Chemical sciences	○○○	○○○	○○○	○○○	○○○	○○○	○○○	○○○	○○○	○○○	○○○	○○○	○○○	○●●
Chemistry	○○	○○	○○	○○		○○		○○		○○				●●
Chemistry with catalysis	○○	○○	○○	○○		○○		○○		○○				●●
Chemistry with materials chemistry	○○	○○	○○	○○		○○		○○		○○				●●
Chemistry with medicinal chemistry	○○	○○	○○	○○		○○		○○		○○				●●
Sheffield														
Chemical physics	●●	●●●												
Chemistry	◑	○	○	○		○	○	○	○	○○○	○	○	●●●	●●●
Southampton														
Chemistry	○	○	○	○		○	○	○		○	○	○	○	○
Strathclyde														
Applied chemistry and chemical engineering	●●	◑		◑		○		○		○	○	○	●●	●
Chemistry	●●	◑		◑		○		○		○	○	○	●●	●
Forensic and analytical chemistry	●●	◑		◑		○		○		○	○	○	●●	●
Sunderland														
Chemical and pharmaceutical science	◑			●		◑ ○		○		○○○				
Surrey														
Chemistry	●●			○	○○○	○	○	○		○○○			●●	●●
Computer-aided chemistry	●●			○	●●●	●●●		○		○○○			●●	●●
Medicinal chemistry	●			●●●	●○○			○○					●●	●●
Sussex														
Chemistry	◑	◑				● ○				○○		○○		
Teesside														
Applied chemistry	●●	●●		●●										●●●
Forensic chemistry	●	●		●	●								●	●

Institution Course title — ● compulsory; ○ optional; ◑ compulsory + options	Supporting and subsidiary content in first/intermediate/final years													
	Mathematics/statistics	Physics	Other physical sciences	Biological science	Computing for chemists	Computer science	History of science	Management	Other business	Foreign language	Social science	Humanities	Communications	Personal skills
UCL														
Chemical physics	●●○	◑◑◑	○	○	●		○			○	○	○	○	●●●
Chemistry	●○○	○○○	○○○	○○○	○○	○	○	○○○		○○○	○	○	○○	●●●
Medicinal chemistry	●○	○	○	●●●	○	○	○	○		○	○	○		●●◑
Warwick														
Biomedical chemistry	○○	○	○○	●●●	●●●	○○		○	○	○○	○	○	●●●	●●●
Chemistry	● ○	○	○	○	●●●	○		○○	○	○○	○	○	●●	●●
Chemistry with medicinal chemistry	● ○	○	○	○	●●●	○	○	○○	○	○○	○	○	●●	●●
York														
Chemistry	●○○	●○○		●○○	●○○	●○○		○○	○○	○○○	○○	○○	●●●	●●●

Table 3c (continued)

Sandwich courses and other industrial experience

Many courses provide a range of opportunities for spending a period of industrial training in the UK or abroad (more information about work and study abroad is given in the next section).

Several institutions offer a choice between a full-time course and a sandwich course: see TABLE 2a. Sandwich courses offer the advantage that after one or more industrial training periods you should be better able to relate theoretical concepts to industrial practice. You will also have a much better feel for the social and cultural aspects of working in industry, and, hopefully, you will have proved that you can work productively in an industrial environment. You may also have made some valuable contacts that will make your job search easier.

On the other hand, sandwich courses take longer than full-time courses. Taking a sandwich course can also affect your social life, as the extra time taken means that you will get 'out of phase' with other students on full-time courses, and during industrial periods you may lose contact with them.

Structure of sandwich courses There are several different ways of organising sandwich courses. There has traditionally been a division between 'thick' and 'thin' sandwich courses, though recently institutions have been moving away from offering thin sandwich courses and there are now comparatively few left. Thick sandwich courses have one or two industrial periods lasting about a year; thin sandwich courses have shorter periods, often of about six months, spent alternately in industry and university or college. Thin sandwich courses can offer the advantage of greater integration of academic work and industrial training, but there are more periods of disruption as you move from one environment to the other.

Thick sandwich courses can be divided into two main types: 2–1–1, in which there is a 12-month period of industrial training between the second and final years of the academic course, and 1–3–1, in which a three-year full-time degree course is 'sandwiched' between two one-year periods in industry.

Institutions usually arrange the periods of industrial experience, but you may be able to arrange your own, subject to approval.

Other forms of industrial experience Some institutions allow students to take an 'intercalated' year out in the middle of a full-time course. In some ways this is similar to a sandwich course, except that you are unlikely to be supervised by the university/college.

TABLE 3d gives information about the opportunities for industrial experience for the courses in this Guide.

Time abroad

Many institutions allow opportunities to spend time abroad in work experience or full-time study. A number of schemes exist for encouraging exchange between students in different countries. One of these is called Erasmus (European Community Action Scheme for the Mobility of University Students), which is part of the wider Socrates scheme.

A number of institutions run named variants of their main chemistry course that include time spent abroad as an integral part of the course. These have names like 'European chemistry' or 'Chemistry with a year in North America'. These courses normally share the same chemistry content as the main chemistry course at that institution, so they are not covered separately in the Guide: see prospectuses for details.

TABLE 3d gives information about the opportunities for industrial experience and spending time abroad while taking the courses in this Guide.

Table 3d Time abroad and sandwich courses

Institution	Course title	Location: ● Europe; ○ North America; ◒ industry; ◑ academic institution	Maximum time abroad (months)	Erasmus available	Sandwich courses: ● thick; ○ thin	Arranged by: ● institution; ○ student	Intercalated industrial year possible
Aberdeen	Chemistry	● ○ ◒ ◑	12	●			
Aston	Applied chemistry				●	● ○	●
	Chemistry	● ◒ ◑	12	●	●	● ○	●
	Chemistry (biotechnology)				●	● ○	
	Chemistry (environmental management)				●	● ○	
	Chemistry (management studies)				●	● ○	
Bangor	Chemistry	● ○ ◒ ◑	12		●	●	●
Bath	*Both courses*	● ○ ◒ ◑	12	●	●	●	●
Belfast	Chemistry		12	●			●
Birmingham	Chemistry	● ◑	12	●	●		●
Bradford	Chemistry with pharmaceutical and forensic science				●	●	●
Brighton	Pharmaceutical and chemical science				●	○	●
Bristol	Chemical physics				●	●	
	Chemistry	● ◒ ◑	9	●	●	●	

(continued) Table 3d

Time abroad and sandwich courses

Institution	Course title	Location: ● Europe	○ North America	◒ industry	◑ academic institution	Maximum time abroad (months)	Erasmus available	Sandwich courses: ● thick	○ thin	Arranged by: ● institution	○ student	Intercalated industrial year possible
Bristol UWE	Biological and medicinal chemistry	●		◒	◑				○		○	●
Cambridge	Natural sciences (chemistry)					3	●					
Cardiff	Chemistry	●		◒	◑	12	●	●		●		●
Dundee	Pharmaceutical chemistry			◒	◑	12	●					●
Durham	Chemistry	●	○	◒	◑	6	●			●	○	●
East Anglia	Biological and medicinal chemistry		○	◒	◑	10		●		●		●
	Chemical physics		○	◒		10		●				●
	Chemistry	●	○	◒	◑	10	●	●		●		●
	Chemistry with analytical science	●	○	◒	◑	10	●	●		●		●
Edinburgh	Chemistry	●	○	◒	◑	12		●		●		●
	Chemistry with environmental chemistry	●	○	◒	◑	12	●	●				●
	Medicinal and biological chemistry					12		●				
Exeter	Biological and medicinal chemistry			◒		12		●		●	○	●
Glamorgan	Chemistry	●		◒		12		●		●	○	●
Glasgow	Chemistry	●	○	◒	◑	12	●	●		●		●
	Chemistry with medicinal chemistry	●		◒	◑	12	●	●		●	○	
Greenwich	*All courses*	●		◒	◑	12	●	●		●		
Heriot-Watt	Chemistry	●	○	◒	◑	12	●	●		●		
	Chemistry with pharmaceutical chemistry	●		◒	◑	6	●					
Huddersfield	*All courses*			◒		12		●		●		
Hull	Chemistry	●		◒	◑	12		●		●	○	●
	Chemistry with analytical chemistry and toxicology	●		◒	◑	12		●		●	○	
	Chemistry with forensic science and toxicology	●		◒	◑	12		●		●	○	●
Imperial College London	Chemistry	●	○	◒	◑	12	●	●			○	
	Chemistry with fine chemicals processing	●	○	◒		12		●			○	
	Chemistry with medicinal chemistry	●	○	◒		12		●			○	
Keele	*Both courses*	●	○	◒	◑	12		●		●	○	●
Leeds	Applied chemistry											●
	Chemistry	●	○	◒	◑	12	●	●		●	○	●
	Chemistry with analytical chemistry	●	○	◒		12		●		●	○	●
	Colour and polymer chemistry											●
	Medicinal chemistry	●	○	◒		12		●		●	○	
Leicester	Chemistry	●	○	◒		9	●	●		●	○	●
	Pharmaceutical chemistry	●	○	◒		9	●					
Liverpool	Chemistry	●		◒		12	●	●		●		●
Liverpool John Moores	*All courses*			◒		12		●		●	○	●
London Metropolitan	Biological and medicinal chemistry	●		◒		12	●	●	○	●	○	●
	Chemistry	●		◒		12		●	○	●	○	●
Loughborough	Chemistry	●	○	◒	◑	12	●	●		●	○	●
	Chemistry with analytical chemistry	●	○	◒		12		●		●	○	●
	Chemistry with forensic analysis							●		●	○	●
	Medicinal and pharmaceutical chemistry	●	○	◒	◑	12	●	●		●	○	●
Manchester	Chemistry	●	○	◒	◑	9	●	●		●	○	●
	Medicinal chemistry					12						
Manchester Metropolitan	Chemical science	●	○	◒	◑	10	●	●		●		●
	Chemistry	●	○		◑	10	●					●

(continued) Table 3d Time abroad and sandwich courses

Institution	Course title	Location: ● Europe; ○ North America; ◒ industry; ◑ academic institution	Maximum time abroad (months)	Erasmus available	Sandwich courses: ● thick; ○ thin	Arranged by: ● institution; ○ student	Intercalated industrial year possible
Newcastle	Chemistry	● ○ ◒ ◑	12	●	●	●	●
	Chemistry with medicinal chemistry	● ◒	12	●	●	●	●
Northumbria	Applied chemistry	● ◒ ◑	12	●	●	●	●
Nottingham	Chemistry	● ○ ◑	12				●
Nottingham Trent	Chemistry	● ○ ◒ ◑	16	●	●	●	●
Paisley	Chemistry	● ◒	15		●	●	●
Plymouth	*Both courses*	○					
Reading	Chemistry	● ◒ ◑	12				●
St Andrews	Chemical sciences				●	●	●
	Chemistry	● ○ ◒ ◑	12	●	●	●	●
	Chemistry with catalysis	● ○ ◒ ◑	12	●	●	●	●
	Chemistry with materials chemistry	● ○ ◒ ◑	12	●	●	●	●
	Chemistry with medicinal chemistry	● ○ ◒ ◑	12	●	●	●	●
Sheffield	Chemistry	◒ ◑	12	●			
Southampton	Chemistry	● ○ ◒ ◑	6		●	●	
Strathclyde	Applied chemistry and chemical engineering	● ○ ◒ ◑	12	●		●	●
	Chemistry	● ○ ◒ ◑	12	●		●	●
	Forensic and analytical chemistry	● ○ ◒ ◑	12	●		●	●
Sunderland	Chemical and pharmaceutical science	● ○ ◒	12	●	○	●	●
Surrey	Chemistry	● ○ ◒ ◑	24		●	●	●
	Computer-aided chemistry	● ◒	12		●	●	
	Medicinal chemistry	● ○ ◒	12		●	●	●
Sussex	Chemistry	● ○ ◒	12		●	●	
Teesside	Applied chemistry	● ◒ ◑	12	●	●	● ○	
	Forensic chemistry	● ◒ ◑	12	●	●	● ○	●
UCL	Chemical physics	●	3	●			
	Chemistry	●	3	●			
	Medicinal chemistry	● ○ ◒ ◑	12	●			
Warwick	Chemistry	● ◒	12	●			
	Chemistry with medicinal chemistry	● ◒	12	●			
York	Chemistry	● ○ ◑	12				●

Most institutions use a variety of assessment methods such as formal written examinations, continuous assessment of coursework and extended projects or dissertations. TABLE 4 gives information about the balance between these methods. It shows in which years there are written examinations and if they contribute to the final degree classification. However, the contribution from examinations in earlier years is often less than the contribution from the examinations in the final year. You should also note that although an examination may not contribute to the final result, passing it may be a condition for continuing with the course.

Many courses allow a wide range of options, which are often assessed in different ways, so it is difficult to give precise figures for the contributions of different assessment methods. For this reason, TABLE 4 shows the possible maximum and minimum contributions from coursework and projects or dissertations.

Projects Projects play an important part in the assessment of many courses. They are usually carried out in the final year and give you the opportunity to pursue particular interests in greater depth, bringing together a range of knowledge and skills learnt during the course.

Projects can be either experimental or, less frequently, non-experimental. Institutions normally provide an extensive list of project suggestions to choose from. You may also be able to suggest your own subject for a project, but this would be subject to approval. The teaching staff will advise you on the choice of project and then supervise the way you carry it out. Projects may be drawn from any part of chemistry. For example, physical/theoretical chemistry projects might include computer-based calculations of the electron distribution in an organic or inorganic molecule, followed by comparison of the results with experimental results. Organic and inorganic chemistry projects might involve the synthesis of a specific target compound, or an investigation of a reaction mechanism. On completing the project, you will have to write a report, and possibly give an oral presentation of the results.

A non-experimental project could be a 15,000-word dissertation on a topic suggested by an academic supervisor, or you may be allowed to work on a topic of your own. Alternatively, you may perform a literature survey, presenting information from a variety of sources on a particular topic. The sources might include academic journals and sophisticated databases – indeed, the result of the project could itself be a database.

Frequency of assessment On many types of course, especially modular courses, assessment is carried out more frequently than in the traditional pattern of end-of-year examinations. Often, each module is assessed independently, soon after it has been completed. The precise details of when assessments are carried out vary from course to course: TABLE 4 shows if courses are assessed every term, semester (there are two semesters in a year) or year. Note that in modular courses, even if all modules are assessed, those occurring early in the course may not contribute as much to the final result as those later in the course.

The mix of assessment methods All assessment systems have advantages and drawbacks: for example, reducing the significance of final examinations may simply mean that short periods of high stress are replaced by a series of deadlines and continuous low-level stress throughout the course. Which of these you prefer will depend on your temperament. Because students vary in their response to different assessment methods, institutions usually employ a combination of methods, which also allows them to match the assessment method to the skill being tested. In some cases you may be able to change the make-up of your assessment regime, for example by choosing a dissertation or project instead of a formal examination.

Table 4 Assessment methods

Institution	**Course title** (Key for frequency of assessment column: ● term; ◑ semester; ○ year)	**Frequency of assessment**	**Years of exams contributing to final degree** (years of exams not contributing to final degree)	**Coursework:** minimum/**maximum** %	**Project/dissertation:** minimum/**maximum** %	**Time spent on projects in final year** %
Aberdeen	Chemistry	◑	(1),(2),(3),**4**	13/**13**	22/**22**	40
	Medicinal chemistry	◑	(1),(2),(3),**4**	13/**13**	22/**22**	40
Aston	Applied chemistry	○	(1),**2,3**	17/**17**	17/**17**	20
	Chemistry	○	(1),**2,3**	11/**11**	17/**17**	20
	Chemistry (biotechnology)	○	**2,3,4**	20/**20**	20/**20**	20
	Chemistry (environmental management)	○	(1),**2,3,4**	20/**20**	25/**25**	20
	Chemistry (management studies)	○	(1),**2,3,4**	20/**20**	20/**20**	20
Bangor	Chemistry	◑	(1),**2,3,4**			33
Bath	Chemistry	◑	(1),**2,3,4**	**10**	**13**	40
	Chemistry for drug discovery	◑	(1),**2,3,4**	**10**	**15**	40
Belfast	Chemistry	◑	(1),**2,3**	6/**6**	13/**13**	20
Birmingham	Chemistry	◑	(1),**2,3,4**		20/**25**	33
Bradford	Chemistry with pharmaceutical and forensic science	◑	(1),**2,3,4**	25/**25**	**8**	25
Brighton	Pharmaceutical and chemical science	◑	(1),**2,3**	24/**30**	24/**26**	30
Bristol	Chemical physics	○	(1),**2,3,4**			45
	Chemistry	○	(1),**2,3,4**	10/**14**	18/**22**	50
Bristol UWE	Biological and medicinal chemistry	◑	(1),**2,3**	40/**50**	10/**30**	35
Cambridge	Natural sciences (chemistry)	○	(1),(2),**3,4**	25/**25**	25/**25**	0
Cardiff	Chemistry	◑	(1),**2,3,4**	22/**22**	13/**13**	17
Dundee	Pharmaceutical chemistry	◑	(1),(2),(3),**4**	**25**	**25**	25
Durham	Chemistry	○	(1),**2,3,4**	20/**25**	5/**10**	10
East Anglia	Biological and medicinal chemistry	○	(1),**2,3,4**	20/**40**	10/**20**	25
	Chemical physics	○	(1),**2,3,4**	20/**40**	10/**20**	25
	Chemistry	○	(1),**2,3,4**	20/**40**	10/**20**	25
	Chemistry with analytical and forensic science	○	(1),**2,3,4**	20/**40**	10/**20**	50
	Chemistry with analytical science	○	(1),**2,3,4**	20/**40**	10/**20**	25
	Pharmaceutical chemistry	○	(1),**2,3,4**	20/**40**	10/**20**	25
Edinburgh	Chemistry	◑	(1),(2),**3,4,5**	0/**6**	23/**28**	50
	Chemistry with environmental chemistry	◑	(1),(2),**3,4,5**	0/**6**	23/**28**	50
	Medicinal and biological chemistry	◑	(1),(2),**3,4,5**	0/**6**	23/**28**	50

(continued) Table 4

Assessment methods

Key for frequency of assessment column: ◐ term; ◑ semester; ○ year

Institution	Course title	Frequency of assessment	Years of exams contributing to final degree (years of exams not contributing to final degree)	Coursework: minimum/**maximum** %	Project/dissertation: minimum/**maximum** %	Time spent on projects in final year %
Exeter	Biological and medicinal chemistry	◑	(1),**2,3,4**	10/**20**	20/**20**	30
Glamorgan	Chemistry	◑	(1),**2,3**	**34**	**20**	20
Glasgow	*Both courses*	◑	(1),(2),**3,4,5**	80/**90**	10/**20**	50
Greenwich	*All courses*	○	(1),**2,3,4,5**	25/**25**	20/**20**	25
Heriot-Watt	*Both courses*	◐	(1),(2),**3,4,5**	5/**5**	20/**20**	50
Huddersfield	*All courses*	◑	(1),**2,3**	30/**40**	12/**16**	25
Hull	Chemistry	◑	**1,2,3,4**	10/**25**	4/**15**	25
	Chemistry with analytical chemistry and toxicology	◑	**1,2,3,4**	10/**25**	4/**7**	25
	Chemistry with forensic science and toxicology	◑	**1,2,3,4**	10/**25**	4/**7**	25
Imperial College London	Chemistry	◑	**1,2,3,4**	21/**21**	13/**13**	75
	Chemistry with conservation science	◑	**1,2,3,4**	21/**21**	13/**13**	45
	Chemistry with fine chemicals processing	◑	**1,2,3,4**	13/**21**	13/**21**	75
	Chemistry with medicinal chemistry	◑	**1,2,3,4**	21/**21**	13/**13**	45
Keele	Chemistry	◑	(1),**2,3,4**	20/**20**	30/**30**	25
	Medicinal chemistry	◑	(1),**2,3**	20/**20**	30/**30**	25
Kingston	Chemistry	◑	(1),**2,3,4**	10/**15**	25/**25**	25
	Chemistry (applied)	◑	(1),**2,3,4**	10/**15**	25/**25**	25
	Medicinal chemistry	◑	(1),**2,3,4**	15	25	
Leeds	Applied chemistry	◑	(1),**2,3**	10/**10**	25/**25**	25
	Chemistry	◑	(1),**2,3,4**	25/**30**	10/**25**	30
	Chemistry with analytical chemistry	◑	(1),**2,3,4**	25/**30**	10/**25**	30
	Colour and polymer chemistry	◑	(1),**2,3**	10/**10**	25/**25**	25
	Medicinal chemistry	◑	(1),**2,3,4**	25/**30**	10/**25**	30
Leicester	Chemistry	◑	**1,2,3,4**	15/**15**	20/**20**	35
	Pharmaceutical chemistry	◑	**1,2,3,4**	15/**15**	20/**20**	35
Liverpool	Chemistry	◑	(1),**2,3,4**	25/**25**	17/**25**	12
	Medicinal chemistry	◑	(1),**2,3**	25/**25**	17/**25**	12
	Medicinal chemistry with pharmacology	◑	(1),**2,3,4**	25/**25**	17/**25**	38
Liverpool John Moores	*All courses*	◑	(1),**2,3,4**	10/**30**	30/**30**	30
London Metropolitan	Biological and medicinal chemistry	◑	(1),**2,3**	30/**50**	15/**15**	25
	Chemistry	◑	(1),**2,3**	30/**50**	15/**15**	50
Loughborough	Chemistry	◑	(1),**2,3,4**	0/**25**	**15**	30
	Chemistry with analytical chemistry	◑	(1),**2,3,4**	0/**25**	**15**	30
	Chemistry with forensic analysis	◑	(1),**2,3,4,5**	0/**25**	**15**	30
	Medicinal and pharmaceutical chemistry	◑	(1),**2,3,4**	0/**25**	**15**	30
Manchester	Chemistry	◑	(1),**2,3,4**	13/**13**	25/**25**	
	Chemistry with patent law	◑	(1),**2,3,4**	13/**13**	25/**25**	
	Medicinal chemistry	◑	**1,2,3,4**	10/**20**	10/**20**	
Manchester Metropolitan	Chemical science	◐	(1),(2),**3,4**	30/**30**	10/**10**	20
	Chemistry	◐	(1),**2,3,4**	30/**30**	10/**10**	20
	Medicinal and biological chemistry	◐	(1),**2,3,4**	30/**30**	10/**10**	20
Newcastle	*Both courses*	◑	(1),**2,3,4**	25/**45**	10/**20**	50
Northumbria	Applied chemistry	◑	(1),**2,4**	20/**30**	20/**30**	25
	Chemistry	○	**3,4**			
Nottingham	Chemistry		(1),**2,3**	7/**10**	7/**10**	
Nottingham Trent	Chemistry	◐	(1),**2,3,4**	40/**45**	26/**26**	30

(continued) Table 4 Assessment methods

Key for frequency of assessment column: ◐ term; ◑ semester; ○ year

Institution	Course title	Frequency of assessment	Years of exams contributing to final degree (years of exams not contributing to final degree)	Coursework: minimum/**maximum** %	Project/dissertation: minimum/**maximum** %	Time spent on projects in final year %
Oxford	Chemistry	○	(1),**2,3,4**	75	25	100
Paisley	Chemistry	◑	(1),(2),**3,4,5**	4/**4**	20/**20**	25
Plymouth	Chemistry (analytical)	◑	(1),**2,3**	40/**40**	22/**22**	25
	Chemistry (applied)	◑	(1),**2,3**	43/**43**	11/**11**	17
Reading	Chemistry	○	(1),**2,3,4**	25/**30**	25/**40**	25
St Andrews	Chemical sciences	◑	(1),(2),**3,4**	**75**	**25**	33
	Chemistry	◑	(1),(2),**3,4**	15/**25**	15/**25**	40
	Chemistry with catalysis	◑	(1),(2),**3,4**	15/**25**	15/**25**	40
	Chemistry with materials chemistry	◑	(1),(2)	15/**25**	15/**25**	40
	Chemistry with medicinal chemistry	◑	(1),(2),**3,4,5**	15/**25**	15/**25**	40
Sheffield	Chemical physics	◑	(1),**2,3,4**	70/**80**	20/**30**	15
	Chemistry	◑	(1),**2,3,4**	88/**96**	4/**12**	33
Southampton	Chemistry	◑	(1),**2,3,4**	12/**14**	15/**28**	33
Strathclyde	Applied chemistry and chemical engineering	◑	(1),(2),(3),**4**		20/**20**	20
	Chemistry	◑	(1),(2),(3),**4,5**		20/**20**	20
	Forensic and analytical chemistry	◑	(1),(2),(3),**4,5**		20/**20**	20
Sunderland	Chemical and pharmaceutical science	◑	(1),**2,3**	30/**30**	15/**15**	
Surrey	Chemistry	◑	(1),**2,3,4**	30/**30**	16/**16**	33
	Computer-aided chemistry	◑	(1),**2**,(3),**4**	30/**30**	16/**16**	30
	Medicinal chemistry	◑	(1),**2,3**	30/**30**	16/**16**	30
Sussex	Chemistry	◐	(1),**2,3,4**	40/**40**	10/**10**	25
Teesside	*Both courses*	◑	(1),**2,3,4**	40/**40**	30/**30**	35
UCL	Chemical physics	○	(1),**2,3,4**	20/**25**	7/**20**	12
	Chemistry	○	(1),**2,3,4**	20/**25**	7/**20**	12
	Medicinal chemistry	○	(1),**2,3,4**	10/**30**	8/**25**	0
Warwick	Biomedical chemistry	○	**1,2,3**	20/**50**	0/**25**	
	Chemistry	○	**1,2,3,4**	27/**30**	7/**10**	20
	Chemistry with medicinal chemistry	○	**1,2,3,4**	27/**30**	6/**10**	20
York	Chemistry	◐	(1),**2,3,4**	12/**20**	8/**18**	10

Chapter 5: Entrance requirements

TABLE 5 lists the entrance requirements for the degree courses in chemistry described in this Guide. You should refer to Chapter 5 in the first part of this Guide for information on how to use the table and how to apply for courses.

The information in TABLE 5 is for general guidance only, since admissions tutors consider applicants individually, and may take many factors into account other than examination grades.

Table 5 Entrance requirements

● compulsory; ○ preferred

Institution	Course title	Number of students (includes other courses)	Typical offers: UCAS points	A-level grades	SCQF Highers grades	A-level Chemistry	A-level Mathematics	A-level Physics	A-level Biology
Aberdeen	Chemistry	(460)	240–260	CDD	BBBB	●			
	Chemistry with e-chemistry		240–260		BBBB				
	Medicinal chemistry	(460)	240–260	CDD	BBBB	●			
Aston	Applied chemistry	20	240–280	BBB	ABBBC	●	○		
	Chemistry	10	240–260	BBB	ABBBC	●	○		
	Chemistry (biotechnology)		240–260	BCC	BBCCC	●	○		
	Chemistry (environmental management)	5	240–260	BCC	BBCCC				
	Chemistry (management studies)		240–260	BCC		●	○		
Bangor	Chemistry	(45)	220–280		BBC	●	○	○	
Bath	Chemistry	80	300–340			●	○		
	Chemistry for drug discovery	20	300–340			●	○		
Belfast	Chemistry	50		CCC	BBBC	●			
	Medicinal chemistry			BBC					
Birmingham	Chemistry	90	280	BBB	BBCCC	●	○	○	○
	Chemistry with analytical science			BBB	BBBCC				
	Chemistry with bio-organic chemistry			BBB	ABBBB				
Bradford	Chemistry with pharmaceutical and forensic science	(75)	260		ABBBB	●	○	○	○
Brighton	Pharmaceutical and chemical science	30	200			○			
Bristol	Chemical physics	10		ABB	AABBB	●	●	●	
	Chemistry	135		BBC	BBBCC	●	○	○	
Bristol UWE	Biological and medicinal chemistry	10	160–200			●			
	Forensic chemistry		240–260			●			
Cambridge	Natural sciences (chemistry)	(600)		AAA		●	○		
Cardiff	Chemistry	85	200–340			●	○	○	
Dundee	Pharmaceutical chemistry	15	240–300			●			
Durham	Chemistry	93		ABB		●	○	○	
East Anglia	Biological and medicinal chemistry	15	300–320	ABB	BBBBB	●	○	○	○
	Chemical physics	10	300–320	ABB	BBBBB	●	●	○	
	Chemistry	35	300–320	ABB	BBBBB	●	○	○	
	Chemistry with analytical and forensic science		320	ABB	AABBB	●	○	○	
	Chemistry with analytical science	10	300	BBB	BBBBB	●	○	○	
	Pharmaceutical chemistry	15	300	BBB	BBBBB	●	○	○	

(continued) **Table 5** Entrance requirements

Institution	**Course title**	**Number of students** (includes other courses)	**Typical offers** UCAS points	A-level grades	SCQF Highers grades	● compulsory; ○ preferred **A-level Chemistry**	**A-level Mathematics**	**A-level Physics**	**A-level Biology**
Edinburgh	Chemical physics	20		BBB	BBBB	●	●	●	
	Chemistry	80		BBB	BBBB	●	○	○	○
	Chemistry with environmental chemistry	10		BBB	BBBB	●	○	○	○
	Medicinal and biological chemistry	30		BBB	BBBB	●	○	○	○
Exeter	Biological and medicinal chemistry	40	280–320			●			●
Glamorgan	Chemistry	25	220–260			●	○	○	
Glasgow	*Both courses*	50		BBC	BBBB	○	○		
Greenwich	Analytical chemistry	20	180			●	○		○
	Chemistry	35	180			●	○		○
	Pharmaceutical chemistry	20	180			●	○		○
Heriot-Watt	Chemistry	55		BCC	BBBC	●	○	○	○
	Chemistry with pharmaceutical chemistry	10		BCC	BBBC	●	○	○	○
Huddersfield	*All courses*	(40)	160–280		BBB	●	○		
Hull	Chemistry	40	240–300			●	○	○	
	Chemistry with analytical chemistry and toxicology	30	240–300			●	○	○	○
	Chemistry with echem		240–300						
	Chemistry with forensic science and toxicology	(90)	240–300			●	○	○	
	Chemistry with molecular medicine		240–300						
	Chemistry with nanotechnology		240–300						
Imperial College London	Chemistry	70		ABB		●	○		
	Chemistry with conservation science	2		ABB		●	○		
	Chemistry with fine chemicals processing	6		ABB		●	○	○	
	Chemistry with medicinal chemistry	18		ABB		●	○		○
Keele	Chemistry	35	220–260		BBCC	●			
	Medicinal chemistry	15	240–260		BBCC	●			
Kent	Forensic chemistry		260	BCC	BBBBC				
Kingston	Chemistry	15	200			●	○		
	Chemistry (applied)	15	200			●	○		
	Medicinal chemistry	15	200			●	○	○	○
Leeds	Applied chemistry	25		BCC		●	○		
	Chemistry	72		BBB		●	○	○	
	Chemistry with analytical chemistry	72		BBB		●	○	○	
	Colour and polymer chemistry	25		BCC		●			
	Medicinal chemistry	72		BBB		●	○	○	
Leicester	Chemical biology		240–300			●			
	Chemistry	70	240–300	CCC	ABBBC	●	○		
	Pharmaceutical chemistry	70	260–300	CCC	ABBBC	●	○		○
Liverpool	Chemistry	(80)	240–280	CCC	BBBBC	●	○	○	
	Medicinal chemistry	(80)	240	CCC	BBCCC	●	○	○	
	Medicinal chemistry with pharmacology	(80)	280	BBC	BBBBC	●	○	○	
Liverpool John Moores	Medicinal and analytical chemistry	(65)	160	CCC		●			○
	Medicinal chemistry	(65)	160	CCC		●			○
	Pharmaceutical science and biological chemistry	(65)	220	BBC		●			○
London Metropolitan	Biological and medicinal chemistry	15	120–160			●			○
	Chemistry	30	120–160			●			

(continued) Table 5 Entrance requirements

Institution	Course title	Number of students (includes other courses)	Typical offers: UCAS points	A-level grades	SCQF Highers grades	● compulsory; ○ preferred: A-level Chemistry	A-level Mathematics	A-level Physics	A-level Biology
Loughborough	Chemistry	30	240–280			●	○		
	Chemistry with analytical chemistry	15	240–280			●	○		
	Chemistry with forensic analysis	25	240–280			●	○		
	Medicinal and pharmaceutical chemistry	30	240–280			●	○		
Manchester	Chemistry	142	220–240	ABC	AABBB	●	○	○	○
	Chemistry with forensic and analytical chemistry			AAB	AAAAB				
	Chemistry with patent law	8	260	AAB	AAAAB	●	○	○	○
	Medicinal chemistry	18	220	ABC	ABBBC	●	○	○	○
Manchester Metropolitan	Analytical chemistry		160–220			●			
	Chemical science	40	100–150			●			
	Chemistry	30	160–220			●	○	○	
	Forensic chemistry		160–220			●			
	Medicinal and biological chemistry	20	160–220			●			○
	Pharmaceutical chemistry		100–150			●			
Newcastle	Chemistry	60		BBC	AABB	●	○		
	Chemistry with medicinal chemistry	30		BBC	AABB	●	○		
Northumbria	Applied chemistry	43	240		CCCCC	●	○		
	Chemistry		260		BBCCC	●			
	Chemistry with forensic chemistry		240		CCCCC				
	Pharmaceutical chemistry		240		CCCCC				
Nottingham	Chemistry	75		BBC		●	○	○	
	Medicinal and biological chemistry			BBB					
Nottingham Trent	Chemistry	80	200–220			●			
	Chemistry with analytical science		200–220						
Oxford	Chemistry	185		AAA	AAAAB	●	○	○	
Paisley	Chemistry	(220)		DD	BBC	●			
	Medicinal chemistry			DD	BBC				
Plymouth	*Both courses*	(60)	260–300			●	○		
Queen Mary	Pharmaceutical chemistry		280						
Reading	Chemistry	50	260–320			●			
	Chemistry with forensic analysis		260–290						
	Chemistry with medicinal chemistry		290–320			●			
St Andrews	Chemical sciences	6		ABB	ABBB	●	○	○	
	Chemistry	60		ABB	ABBB	●	○	○	
	Chemistry with catalysis	5		ABB	ABBB	●	○	○	
	Chemistry with materials chemistry	5		ABB	ABBB	●	○	○	
	Chemistry with medicinal chemistry	10		ABB	ABBB	●	○	○	
Sheffield	Chemical physics	10		BBB	BBBB	●	●	●	
	Chemistry	140		BCC	BBBB	●	○	○	
Southampton	Chemistry	(80)	320	BBC		●			
Strathclyde	Applied chemistry and chemical engineering			ABB	AAAC	●	●	○	○
	Chemistry	50		BBC	BBBB	●	●	○	○
	Chemistry with drug discovery			ABB	AAAC				
	Forensic and analytical chemistry	50		ABB	AAAC	●	●	○	○
Sunderland	Chemical and pharmaceutical science	50	240–360		BBBB	●	○		
	Chemistry		200–360		BBCC				

(continued) Table 5 Entrance requirements

Institution	Course title	Number of students (includes other courses)	Typical offers: UCAS points	A-level grades	SCQF Highers grades	● compulsory; ○ preferred: A-level Chemistry	A-level Mathematics	A-level Physics	A-level Biology
Surrey	Chemistry	50	240–260	BCC		●	○	○	
	Chemistry with forensic investigation		260			●			
	Computer-aided chemistry	10	260			●	○	○	
	Medicinal chemistry	10	260			●	○	○	○
Sussex	Chemistry	(110)		ABB	AABBB	●			
	Chemistry with forensic science			BBB	ABBB				
Teesside	Applied chemistry	20	180–240			○	○	○	
	Forensic chemistry	20	140–200			○	○	○	
UCL	Chemical physics	10		ABB		●	●	●	
	Chemistry	(80)		ABB		●	○	○	○
	Medicinal chemistry			ABB		●	○	○	○
Warwick	Biomedical chemistry	40	320–350			●	○		○
	Chemistry	50	320–350			●	○	○	
	Chemistry with medicinal chemistry	20	320–350			●	○		○
York	Chemistry	110		BBB	BBBBB	●	○	○	
	Chemistry, biological and medicinal chemistry			BBB	BBBBB	●			

Chapter 6: Professional organisations

The Royal Society of Chemistry is the professional body for chemists. It is the largest organisation in Europe for advancing the chemical sciences, with around 45,000 members. RSC activities span education and training, conferences and science policy, and the promotion of the chemical sciences to the public.

The RSC offers a wide variety of support for students considering a degree in the chemical sciences, including advice on choosing between courses and institutions, and information on career opportunities. It also recognises and gives accreditation to degree courses. For more information contact the Education Department at the RSC, Burlington House, Piccadilly, London W1J 0BA, www.rsc.org/education.

If you want to practise in the National Health Service as a biomedical scientist, you must be registered by the Health Professions Council: see the *Biological Sciences* Guide for more details.

Chapter 7: Further information

Background information Chapter 7 of the first part of this Guide contains a list of publications giving information on admission to higher education, financial support and access for special groups of applicants.

Background reading in chemistry

Molecules P W Atkins. CUP, 2003 (2nd edition), £22.99
The Periodic Kingdom P W Atkins. Orion, 1996, £6.99
The Consumer's Good Chemical Guide: A Jargon-Free Guide to Controversial Chemicals J Emsley. OUP, 1998, £9.99
The Crumbs of Creation: Trace Elements in History, Medicine, Industry, Crime and Folklore J Lenihan. Institute of Physics, 1988, £23.99
Chemistry: Molecules, Matter and Change P W Atkins & L Jones. W H Freeman, 2000 (4th edition), £35.99
Marvels of the Molecule L Salem. Wiley, 1987, £16.95
Asimov's New Guide to Science I Asimov. Penguin, 1993, £15.99
The Double Helix J Watson. Penguin, 1999 (2nd edition), £8.99
The Chemistry of Life S Rose & R Mileusnic. Penguin, 1999 (4th edition), £12.99
A Primer to Mechanism in Organic Chemistry P Sykes. Pearson Higher Education, 1995, £25.99
Foundations of Organic Chemistry M Hornby & J Peach. OUP, 1993, £10.99
Essentials of Inorganic Chemistry I D M P Mingos. OUP, 1995, £10.99
Foundations of Physical Chemistry C Lawrence, A Rodger & R Compton. OUP, 1996, £8.99

The courses This Guide gives you information to help you narrow down your choice of courses. Your next step is to find out more about the courses that particularly interest you. Prospectuses cover many of the aspects you are most likely to want to know about, but some departments produce their own publications giving more specific details of their courses. University and college websites are shown in TABLE 2a.

You can also write to the contacts listed below.

Aberdeen Student Recruitment and Admissions Service (sras@abdn.ac.uk), University of Aberdeen, Regent Walk, Aberdeen AB24 3FX

Aston Admissions Tutor, Chemical Engineering and Applied Chemistry, Aston University, Aston Triangle, Birmingham B4 7ET

Bangor Dr I R Butler, Admissions Tutor, Department of Chemistry, University of Wales, Bangor, Gwynedd LL57 2UW

Bath Dr S J Roser (s.j.roser@bath.ac.uk), School of Chemistry, University of Bath, Bath BA2 7AY

Belfast Dr J F Malone, School of Chemistry, The Queen's University of Belfast, Belfast BT9 5AG

Birmingham Dr Ian Gameson (ugadmissions@chemistry.bham.ac.uk), Admissions Co-ordinator, School of Chemistry, University of Birmingham, Birmingham B15 2TT

Bradford Dr Steve Dobrowski (cfs.admissions@bradford.ac.uk), Director of Forensic Sciences, Department of Chemical and Forensic Sciences, University of Bradford, Bradford BD7 1DP

Brighton Dr D Naughton, School of Pharmacy and Biomolecular Sciences, University of Brighton, Moulsecoomb, Brighton BN2 4GJ

Bristol Chemical physics Mrs D Roberts, Undergraduate Admissions Secretary, School of Chemical Physics; Chemistry Dr D W Thompson (d.w.thompson@bristol.ac.uk), School of Chemistry; both at University of Bristol, Cantock's Close, Bristol BS8 1TS

Bristol UWE Dr D Thornton (dilys.thornton@uwe.ac.uk), Faculty of Applied Sciences, University of the West of England, Coldharbour Lane, Bristol BS16 1QY

Cambridge Cambridge Admissions Office (admissions@cam.ac.uk), University of Cambridge, Fitzwilliam House, 32 Trumpington Street, Cambridge CB2 1QY

Cardiff Dr P C Griffiths (chemistry-ug@cardiff.ac.uk), Department of Chemistry, Cardiff University, PO Box 912, Cardiff CF10 3TB

Dundee Admissions and Student Recruitment (srs@dundee.ac.uk), University of Dundee, Dundee DD1 4HN

Durham Dr A K Hughes (a.k.hughes@durham.ac.uk), Chemistry Department, University of Durham, Durham DH1 3LE

East Anglia Admissions Officer (che.admiss@uea.ac.uk), School of Chemical Sciences, University of East Anglia, Norwich NR4 7TJ

Edinburgh Undergraduate Admissions Office (sciengug@ed.ac.uk), College of Science and Engineering, University of Edinburgh, The King's Buildings, West Mains Road, Edinburgh EH9 3JY

Exeter Dr Steve Aves (s.j.aves@exeter.ac.uk), Washington Singer Building, University of Exeter, Perry Road, Prince of Wales Road, Exeter EX4 4XB

Glamorgan Dr Wynne Evans, School of Applied Sciences, University of Glamorgan, Pontypridd, Mid Glamorgan CF37 1DL

Glasgow Dr R Hill (r.hill@chem.gla.ac.uk), Department of Chemistry, Glasgow University, Glasgow G12 8QQ

Greenwich Dr M J K Thomas (tm04@gre.ac.uk), School of Science, University of Greenwich, Central Avenue, Chatham Maritime ME4 4TB

Heriot-Watt Dr J E Parker (j.e.parker@hw.ac.uk), Department of Chemistry, Heriot-Watt University, Riccarton, Edinburgh EH14 4AS

Huddersfield Dr C R Rice (c.r.rice@hud.ac.uk), Admissions Tutor, Department of Chemical and Biological Sciences, University of Huddersfield, Queensgate, Huddersfield HD1 3DH

Hull Dr M Hird, School of Chemistry, University of Hull, Hull HU6 7RX

Imperial College London Dr Edward H Smith (ed.smith@imperial.ac.uk), Department of Chemistry, Imperial College London, South Kensington Campus, London SW7 2AZ

Keele Dr A Curtis (a.d.m.curtis@chem.keele.ac.uk), Admissions Tutor, Department of Chemistry, Keele University, Staffordshire ST5 5BG

Kent Registry (recruitment@kent.ac.uk), University of Kent, Canterbury, Kent CT2 7NZ

Kingston Student Information and Advice Centre, Cooper House, Kingston University, 40–46 Surbiton Road, Kingston upon Thames KT1 2HX

Leeds Applied chemistry Colour and polymer chemistry Dr Mark Heron (b.m.heron@leeds.ac.uk), Department of Colour Chemistry; All other courses Michelle Lesniasnski (admissions@chem.leeds.ac.uk), School of Chemistry; both at University of Leeds, Leeds LS2 9JT

Leicester Dr Paul Jenkins (chemadmiss@le.ac.uk), Department of Chemistry, University of Leicester, Leicester LE1 7RH

Liverpool Ms Samantha Dunn (ucas@ch.liv.ac.uk), Undergraduate Recruitment Office, Department of Chemistry, University of Liverpool, Crown Street, Liverpool L69 7ZD

Liverpool John Moores Student Recruitment Team (recruitment@ljmu.ac.uk), Liverpool John Moores University, Roscoe Court, 4 Rodney Street, Liverpool L1 2TZ

London Metropolitan Admissions Office (admissions@londonmet.ac.uk), London Metropolitan University, 166–220 Holloway Road, London N7 8DB

Loughborough Dr R J Mortimer (r.j.mortimer@lboro.ac.uk), Department of Chemistry, Loughborough University, Loughborough LE11 3TU

Manchester Mrs Mary Lea (mlea@manchester.ac.uk), School of Chemistry, University of Manchester, Manchester M13 9PL

Manchester Metropolitan Dr Ken Williamson, Admissions Tutor, Chemistry and Materials Department, Manchester Metropolitan University, Chester Street, Manchester M1 5GD

Newcastle Dr C Bleasdale (chemistry.ugadmin@ncl.ac.uk), School of Natural Sciences (Chemistry), University of Newcastle upon Tyne, Bedson Building, Newcastle upon Tyne NE1 7RU

Northumbria Admissions (er.educationliaison@northumbria.ac.uk), University of Northumbria, Trinity Building, Northumberland Road, Newcastle upon Tyne NE1 8ST

Nottingham Dr P Hubberstey (peter.hubberstey@nottingham.ac.uk), School of Chemistry, University of Nottingham, University Park, Nottingham NG7 2RD

Nottingham Trent Admissions Office (sci.enquiries@ntu.ac.uk), School of Biomedical and Natural Sciences, Nottingham Trent University, Clifton, Nottingham NG11 8NS

Oxford Nina Jupp (nina.jupp@chem.ox.ac.uk), Central Chemistry Laboratory, Oxford University, South Parks Road, Oxford OX1 3QH

Paisley Head of Department of Chemistry and Chemical Engineering, University of Paisley, High Street, Paisley PA1 2BE

Plymouth Faculty of Science, University of Plymouth, Drake Circus, Plymouth PL4 8AA

Queen Mary Admissions and Recruitment office (admissions@qmul.ac.uk), Queen Mary University of London, Mile End Road, London E1 4NS

Reading Dr Matthew Almond (chemistry@rdg.ac.uk), Department of Chemistry, University of Reading, Whiteknights, Reading RG6 6AD

St Andrews Dr David Richens (dtr@st-and.ac.uk), School of Chemistry Admissions, University of St Andrews, North Hough, St Andrews KY16 9ST

Sheffield Chemical physics Dr J P A Fairclough (p.fairclough@sheffield.ac.uk); Chemistry Senior Tutor for Admissions; both at Department of Chemistry, University of Sheffield, Sheffield S3 7HF

Southampton Dr J D Hinks (j.d.hinks@soton.ac.uk), Admissions Tutor, School of Chemistry, University of Southampton, Southampton SO17 1BJ

Strathclyde Dr D Pugh (d.pugh@strath.ac.uk), Associate Chairman, Department of Pure and Applied Chemistry, University of Strathclyde, 295 Cathedral Street, Glasgow G1 1XL

Sunderland Student Recruitment (student-helpline@sunderland.ac.uk), University of Sunderland, Chester Road, Sunderland SR1 3SD

Surrey Dr Bredan Howlin, Chemistry (C4), SBMS, University of Surrey, Guildford GU2 7XH

Sussex Admissions Tutor (ug.admissions@chemistry.sussex.ac.uk), Department of Chemistry, University of Sussex, Falmer, Brighton BN1 9QJ

Teesside Sandra Joyce, University of Teesside, Middlesbrough, Cleveland TS1 3BR

UCL Chemical physics Chemistry Dr D A Tocher; Medicinal chemistry Dr A B Tabor; both at Department of Chemistry, University College London, 20 Gordon Street, London WC1H 0AJ

Warwick Admissions Officer (s.pearson@warwick.ac.uk), Department of Chemistry, University of Warwick, Coventry CV4 7AL

York Dr R J Mawby, Department of Chemistry, University of York, York YO10 5DD

The CRAC Series of Degree Course Guides

Series 1

Architecture, Planning and Surveying
Business, Management and Economics
Classics, Theology and Religious Studies
Engineering
English, Media Studies and American Studies
Hospitality, Tourism, Leisure and Sport
Mathematics, Statistics and Computer Science
Music, Drama and Dance
Physics and Chemistry
Sociology, Anthropology, Social Policy and Social Work

Series 2

Agricultural Sciences and Food Science and Technology
Art and Design Studies and History of Art and Design
Biological Sciences
Environmental Sciences
Geography and Geological Sciences
History, Archaeology and Politics
Law and Accountancy
Medical and Related Professions
Modern Languages and European Studies
Psychology, Philosophy and Linguistics

Trotman Editorial and Publishing Team

Editorial Mina Patria, Editorial Director; Jo Jacomb, Editorial Manager; Ian Turner, Production Editor
Production Ken Ruskin, Head of Manufacturing and Logistics; James Rudge, Production Artworker
Advertising and Sales Tom Lee, Commercial Director; Sarah Talbot, Advertising Manager
Sales and Marketing Sarah Lidster, Marketing Manager
Cover design XAB and James Rudge